Short-Run SPC for Manufacturing and Quality Professionals

The principle of "value comes from the production of parts rather than charts" is of vital importance on the shop floor when using practical statistical process control (SPC). The production worker should need to enter only a sample's measurements to get immediately actionable information as to whether corrective action (e.g., as defined by a control plan's reaction plan) is necessary for an out-of-control situation, and should not have to perform any calculations, draw control charts, or use sophisticated statistical software. This book's key benefit for the readers consists of spreadsheet deployable solutions.

Traditional SPC relies on the assumption that sufficient data are available with which to estimate the process parameters and set suitable control limits. Many practical applications involve, however, short production runs for which no process history is available. There are nonetheless tested and practical control methods such as PRE-Control and short-run SPC that use the product specifications to set appropriate limits. PRE-Control relies solely on the specification limits while short-run SPC starts with the assumption that the process is capable—that is, at least a 4-sigma process, and works from there to set control limits. Cumulative Sum (CUSUM) and exponentially weighted moving average (EWMA) charts also can be used for this purpose. Specialized charts can also track multiple part characteristics, and parts with different specifications, simultaneously. This is often useful, for example, where the same tool is engaged in mixed-model production.

Readers will be able to deploy practical and simple control charts for production runs for which no prior history is available and control the processes until enough data accumulate to enable the traditional methods (assuming it ever does). They will be able to track multiple product features with different specifications and also control mixed-model applications in which a tool generates very short runs of parts with different specifications. The methods will not require software beyond readily available spreadsheets, nor will they require specialized tables that are not widely available. Process owners and quality engineers will be able to perform all supporting calculations in Microsoft Excel, and without the need for advanced software.

Short-Run SPC for Manufacturing and Quality Professionals

William A. Levinson

Routledge
Taylor & Francis Group

A PRODUCTIVITY PRESS BOOK

First published 2023
by Routledge
605 Third Avenue, New York, NY 10158

and by Routledge
4 Park Square, Milton Park, Abingdon, Oxon, OX14 4RN

Routledge is an imprint of the Taylor & Francis Group, an informa business

ISBN: 9781032249902 (hbk)
ISBN: 9781032249896 (pbk)
ISBN: 9781003281061 (ebk)

DOI: 10.4324/9781003281061

Typeset in Garamond
by Deanta Global Publishing Services, Chennai, India

Contents

Preface

The musical comedy *Guys and Dolls* features a scene in which Big Jule, a gangster from Chicago, wants to shoot dice against Nathan Detroit (played by Frank Sinatra). Detroit objects that the dice have no spots on them only to have Big Jule reassure him that he remembers where the spots formerly were. A similar situation arises when we start up a new production run that has no prior history from which we can estimate its mean and standard deviation. If we paraphrase the exchange between Nathan Detroit and Big Jule, it might go like this:

> But we have no information about this process' mean or standard deviation.
> That's a piece of bad luck, but I can set up control charts nonetheless.
> You are going to set up control charts without any process history?

The Automotive Industry Action Group's (2016) CQI-26 *SPC Short Run Supplement* tells us how to do exactly that, and this book will show how to deploy these solutions to the shop floor in Microsoft Excel. This makes it unnecessary to purchase a sophisticated software package that might not even be designed for use on the shop floor, and there appear to be few available that even address the more advanced applications of short-run SPC. We can alternatively use PRE-Control, which relies solely on the specification limits and also cumulative sum (CUSUM), which will detect relatively small discrepancies between the process mean and the nominal.

Additional complications arise in mixed-model production as opposed to long production runs. If, for example, a downstream process' bill of materials consists of two units of A, one of B, and one of C, then the ideal production pattern is 2A, 1B, and 1C as opposed to long production runs of each.

Suppose the same tool can produce all of them, which again is the ideal, but they have different specifications. Short-run SPC charts can allow the tool's performance to be tracked easily on a single control chart. Yet another complication involves parts with multiple features, again as produced by a single tool that performs multiple operations on the part, each with its own specification limits and possibly standard deviation. Short-run SPC charts can address this issue as well.

A principal consideration is the ease with which the methods can actually be deployed on the shop floor. Time spent on data entry by production workers does not add value so it should be kept to a minimum or, to put it more simply, value comes from production of parts rather than charts. As Robert A. Heinlein put it in *Starship Troopers* (1959), "If you load a mud foot down with a lot of gadgets that he has to watch, somebody a lot more simply equipped—say with a stone ax—will sneak up and bash his head in while he is trying to read a vernier." This book's key deliverable is something with all the mathematical precision of the vernier along with the simplicity of a stone ax. The production worker should ideally need to do no more than enter the measurements on a spreadsheet, which then performs all the necessary calculations to return a simple signal (such as a cell highlighted in red) that indicates whether the process is out of control. The process owner or quality engineer may have to program the spreadsheet to do this, and this book provides guidance and examples, but the worker's interface should be as simple as possible. The book will have fulfilled its purpose if it puts practical and effective short-run SPC within easy reach of process owners and production workers.

William A. Levinson, P.E., is the principal of Levinson Productivity Systems, P.C. He is an ASQ fellow, certified quality engineer, quality auditor, quality manager, reliability engineer, and Six Sigma Black Belt holder. He holds degrees in chemistry and chemical engineering from Penn State and Cornell Universities, and night school degrees in business administration and applied statistics from Union College, and he has given presentations at the ASQ World Conference, TOC World 2004, and other national conferences on productivity and quality.

Levinson is also the author of *The Expanded and Annotated My Life and Work: Henry Ford's Universal Code for World-Class Success. Henry Ford's Lean Vision* is a comprehensive overview of the lean manufacturing and organizational management methods that Ford employed to achieve unprecedented bottom-line results, and *Beyond the Theory of Constraints* describes how Ford's elimination of variation from material transfer and processing times allowed him to come close to running a balanced factory at full capacity. *Statistical Process Control for Real-World Applications* shows what to do when the process doesn't conform to the traditional bell curve assumption.

Education

B.S. Pennsylvania State University

M.Eng. Cornell University

MBA Union College (night school)

M.S. in Operations Research and Applied Statistics (Union College, night school)

Introduction

Control plans for manufacturing processes often include statistical process control (SPC) along with a reaction plan that tells the production worker how to correct an undesirable change in the process' mean or variation. Quality and manufacturing professionals are most familiar with traditional or "textbook" SPC, and statistical software packages can rapidly estimate the process parameters with which to define the control limits. They can also generate a control chart with which to test the assumption that the data come from an in-control process, i.e. one in which there are no special or assignable causes that might overestimate the variation.

Statistical software can also test the traditional assumption that the data come from a normal (bell curve) distribution, and this is important because bell curves are far more common in textbooks than they are in factories. If this assumption is not met, however, StatGraphics, just one example, can estimate the parameters of non-normal distributions and set appropriate control limits for them. This can even be done with Excel's built-in Solver feature along with the maximum likelihood estimation (MLE) method. The same goes for related process capability and process performance studies that assess a process' ability to meet specifications, and also expose long-term variation that is not reflected in the sampling plan.

All these assume, however, that there are enough historical data for the statistical software to use. This is not the case when we start up a new process or, in many cases, make a changeover at a workstation. A change-over is technically a new process because any change in a process factor (manpower, machine, material, measurement, method, or environment) introduces the risk of unforeseen and potentially undesirable consequences. Short production runs and mixed-model protection involve a lot of change-overs by design, and this makes short-run SPC a vital supporting technique.

The AIAG reference for short-run SPC (2016) offers a simple way to address these situations. We can estimate the process parameters from similar processes, or past histories from the same process. We can alternatively assume that the process is capable, which means usually that there are no fewer than four process standard deviations ("sigmas") between the nominal and each specification limit. This gives us the estimation of process variation we need to define control limits. If the process turns out to not be capable, then we will eventually get points above the upper control limit for the range (R) or sample standard deviation (s) chart for process variation. We can also detect much more rapidly, especially with the cumulative sum (CUSUM) and exponentially weighted moving average (EWMA) methods, discrepancies between the process mean and the target nominal.

Matters become even more complicated when a tool makes different parts. There was a time when, if a product's bill of material consisted of 3A, 1B, and 2C, all of which can be made by the same tool, we might make 3000 As, 1000 Bs, and 2000 Cs and put them into inventory for use as needed. A contemporary small lot production scheme or kanban container might, in contrast, require 3As, 1B, and 2Cs. Even if these have different nominals, however, it is very easy to standardize the sample statistics by converting them into their standard normal deviates (z) for display on a single control chart. A spreadsheet cell can change color to indicate a result outside the control limits, so a chart might not even be needed except for information purposes.

A similar issue involves multiple features on a part, all of which are created by the same tool. The AIAG reference (2016, p. 25) has an example in which a drill makes holes of three different sizes in a valve body for hydraulic control. We don't really want three charts for one tool but Wise and Fair (1998, Chapter 19) describe the *group chart* that can track multiple product features. These need not have the same specifications, nominals, or even standard deviations. These are deployable in Excel, and it is again not even necessary to make a chart. Spreadsheet cells can be programmed to change colors when certain conditions are exceeded to give the production worker an immediate signal to this effect.

This book will address the shop floor's options as follows.

Chapter 1 will cover PRE-Control which will work for applications with one-sided as well as two-sided specifications. PRE-Control is surprisingly robust to distributional assumptions and can be put to work for the first item that comes out of a tool or workstation.

Chapter 2 will introduce the basics of short-run SPC chart deployment given limited or no prior knowledge of the process' parameters. The only requirement is knowledge of the specification limits along with the assumption that the critical to quality (CTQ) characteristic follows the normal distribution. This is because we assume the process mean to be the nominal, which is usually half way between the specification limits. We also infer the standard deviation from the required process performance index. This approach does not, however, carry over into two-parameter non-normal distributions such as the gamma and Weibull distributions.

Chapter 3 will show how to deploy CUSUM (cumulative sum) and exponentially weighted moving average (EWMA) charts in a spreadsheet format that can detect small changes in the process mean. These are useful for startup processes, as well as any process in which the objective is to detect small rather than large shifts. If, for example, a setup's mean deviates from the nominal by half a process standard deviation, a CUSUM chart for individuals is likely to detect this far more quickly than a traditional Shewhart chart. This chapter also provides awareness of the extremely powerful Cuscore chart whose purpose is to detect *specific* assignable or special causes such as an increase in tool wear rate.

Chapter 4 will address situations in which the process output consists of parts with different nominals, or multiple characteristics which, however, come from the same tool.

Chapter 5 will provide an overview of the acceptance control chart, whose purpose is to control processes in which it is not practical and/ or desirable to hold the process mean to the nominal.

Chapter 1

PRE-Control

PRE-Control (Juran and Gryna, 1988, 24.31 to 24.38) is intended specifically for qualification of new processes. PRE-Control is emphatically not statistical control because it makes no assumptions about the distribution of the critical to quality (CTQ) characteristic. It is the simplest approach to setup and monitoring of processes for which there is no prior history whatsoever, and it requires knowledge only of the specification limits.

AIAG (2005, 104) adds that PRE-Control is a form of "stoplight control" and is emphatically not process control; it is instead nonconformance control. It assumes that the combined process and measurement system variation are less than or equal to the specification width, i.e. the process is a 3-sigma process or better. Another way of saying this is that 99.73% or more of the work will be within specification. These assumptions apply to the "nominal is best" model, noting that the other models are for one-sided specification limits. The same reference adds an additional assumption, namely, a flat loss function as opposed to, for example, Taguchi's quadratic loss function. Parts inside the specification limits are assumed to be equally good, and those outside of them equally bad.

Process qualification requires successful production of five consecutive parts in a specified zone that adjoins or surrounds the process nominal. Process monitoring consists of measurement or inspection of pairs of items on what we can think of as a "two fouls or one strike and you're out" basis. Fouls consist of measurements in the yellow zone, which is set at a certain distance from the nominal. A strike consists of a measurement in the red zone, which is out of specification. The prerequisite for PRE-Control is therefore only that the parts be measurable with an instrument or gage that

DOI: 10.4324/9781003281061-1

returns real number data, or appraisable with go/no-go gages that can be set (for example, with gage blocks) at specified distances from the nominal.

Two-Sided Tolerance: Nominal Is Best

Given a bilateral specification limit LSL (lower specification limit) and USL (upper specification limit), with the nominal halfway between them, define the zones as follows. Red is anything outside the specification limits, as shown in Figure 1.1. This figure can be rotated 90 degrees clockwise for horizontal display as well, with the LSL on the left and the USL on the right.

$$\text{Green} = \text{nominal} \pm \frac{\text{USL} - \text{LSL}}{4}$$

$$\text{Yellow} = \text{nominal} \pm \frac{\text{USL} - \text{LSL}}{2} \text{exclusive of green}$$

The yellow zone is alternatively everything between the green zone and the specification limit, while the green zone is half the distance from the nominal to each specification limit.

Suppose, for example, that a feature diameter has a nominal size of 0.50 ± 0.02 inch.

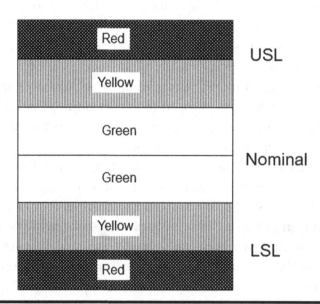

Figure 1.1   PRE-Control, two-sided tolerance

$$\text{Green} = 0.5 \pm \frac{0.52 - 0.48}{4} = [0.49, 0.51]$$

$$\text{Yellow} = 0.5 \pm \frac{0.52 - 0.48}{2} = [0.48, 0.52] \text{ exclusive of green}$$

The result is as shown in Figure 1.2, which can alternatively be displayed horizontally.

Process Qualification

The first step is to qualify the process by making parts until five consecutive ones fall within the green zone. Juran and Gryna (1988, 24.32) stress that this is not process control; its purpose is to ensure that the process capability is sufficient to deliver consistently in-specification product. If it is not possible to get five consecutive measurements inside the green zone, then it is necessary to identify the root cause because the process will not be able to make consistently good product. Production can begin once the process has been qualified.

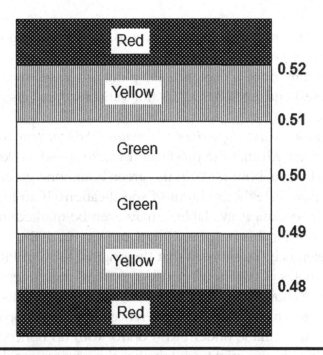

Figure 1.2   PRE-Control zones for specification 0.50 ± 0.02

Process Monitoring

Process monitoring consists of periodic appraisal of samples of two items, A and B. If both fall into the yellow zone above or below the nominal, this is evidence that the process mean has changed and that adjustment is necessary. If one falls into a red zone and is out of specification, this also is evidence of trouble that requires corrective action to avoid generation of additional nonconforming product. If, however, A and B fall into yellow zones on opposite sides of the nominal, this is evidence that the variation is too large. That is,

- A and B are in the green zone; no action is required. If the first piece is in the green zone, there may in fact be no need to check the second (Smith, 2009).
- A or B is in the green zone and the other part is in the yellow zone; no action is required.
- A and B are in the yellow zone on the same side of the nominal; assume that the process mean has shifted and corrective action is required.
- A and B are in opposite yellow zones; conclude that the variation has increased.

If adjustment is required, then the process must be re-qualified with five consecutive parts in the green zone. This is a very simple variant of switching rules that are common for acceptance sampling processes such as ANSI/ASQ Z1.4 (for attributes) and ANSI/ASQ Z1.9 (for variables), and also continuous sampling plans.

The process monitoring stage does not require the construction of an actual control chart. All that the production worker needs to know is whether he or she gets both parts in the green zone, one green and one yellow, both yellow, or either red (out of specification). If an instrument that returns real number data is available, it may even be marked in green, yellow, and red zones for this purpose (Juran and Gryna, 1988, 24.34).

The same reference (24.33) adds that the original form of PRE-Control called for 25 samples of two items between process adjustments but more recent experience shows that six samples between adjustments will deliver almost no nonconforming work. The reference cites an example in which 30 million parts were made under PRE-Control with no nonconformances whatsoever, although one might ask whether the process capability was extremely high because random variation from even a centered 4-sigma process will generate 63.3 nonconformances per million opportunities.

Go/No-Go Gages

Suppose instead that go/no-go gages are available that can be adjusted to specific sizes (e.g. with gage blocks) or come in specific sizes (such as plug gages). Assume, for example, that a nominal hole diameter is 0.50 inches, with specification ±0.02 inches. It might be possible to address this with a plug gage whose smaller ("go") end is 0.49 inch and whose larger ("no-go") end is 0.51 inch. If the small end fits the hole and the large end does not, then the diameter is between 0.49 and 0.51 inches which puts the part in the green zone. If the small end does not fit the hole, then the diameter is less than 0.49, and the inspector would try again with a 0.48 inch gage to determine whether the part is in the yellow zone or is actually nonconforming. If the large end fits the hole, then it is 0.51 inch or larger, and the inspector would try again with a 0.52 inch gage. The same approach could be extended to other go/no-go gages as well. Feigenbaum (1991, 449-450) says explicitly that PRE-Control is very amenable to snap ages and other go/no-go gages.

Producer's and Consumer's Risks

The next step is to quantify the producer's (false alarm) and consumer's (failure to detect an assignable cause) risk for a PRE-Control chart. Montgomery (1991, 333) elaborates that these rules assume that a nonconforming fraction of 1–3% is acceptable and also that the process capability index is at least 1.15, which corresponds to a 3.45-sigma process. This means the process is only marginally capable because the usual rule is that 4-sigma is the minimum to qualify as "capable."

If we assume instead a 3-sigma process (capability index = 1), which definitely falls short of the definition of capable, we can compute the false alarm risks. Figure 1.3 from StatGraphics shows that, given a lower specification of 0 and an upper specification of 1, the chance of a point falling outside the green zone is 0.1337.

This is, however, a two-sided risk, so the risk that a point will be in the lower zone is 0.0669. The chance that two consecutive points will be there is only 0.00447. We must double this to account for the chance that two consecutive points will be in the upper yellow zone to get 0.00894 or about

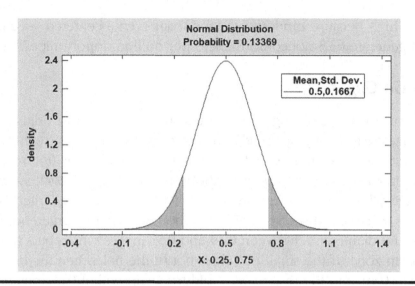

Figure 1.3 Chance of a yellow zone result if the process is centered

0.89%. This is the Type I or producer's risk that PRE-Control will tell us that the process mean has shifted when it hasn't.

We must then add the false alarm risk of A being in the upper yellow zone and B being in the lower, and the alternate combination of A low and B high. This adds a 0.89% risk that PRE-Control will tell us that the variation has increased when it hasn't. The overall false alarm risk is therefore 0.0179 (1.79%); Juran and Gryna get 4/196 (0.0204) using one-sided risks of 1/14. The same reference adds that the worst case consumer's situation arises for a 3-sigma process whose mean has shifted 0.85 process standard deviations, in which case the nonconforming fraction will not exceed an average of 1%.

One-Sided Specification Limits

PRE-Control can also address applications with one-sided specification limits such as flatness or concentricity (Juran and Gryna, 1988, 24.33) and presumably trace contaminants and trace impurities in materials. There is no lower specification limit because we do not care how little we get, and it is impossible to get less than zero. This is particularly important because short-run SPC methods are not usable on most non-normal distributions that characterize these processes in the absence of prior history.

Short-run SPC can start with the assumption that the process is capable, i.e. there are four standard deviations between the nominal and each

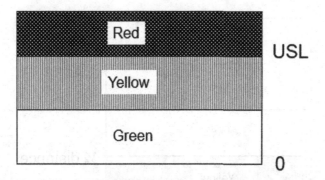

Figure 1.4   PRE-Control, total indicator reading, one-sided specification limit

specification limit, and work from there to develop usable control charts. We can assume similarly that a gamma distribution, which is often a good model for trace contaminants and impurities, is capable and will deliver no more than 31.7 nonconformances per million opportunities at its upper specification limit. This is the same nonconforming fraction that a four-sigma process delivers at each specification limit. The problem is that the gamma distribution has an almost limitless combination of shape and scale parameters that will meet this requirement, and addition of a threshold parameter* can make the situation even more complicated. This leaves PRE-Control as essentially the only option with which to qualify and monitor the process in question.

The PRE-Control system for a one-sided specification is, however, simpler than that for a two-sided one. Place the boundary of the green and yellow zone halfway between zero and the upper specification limit as shown in Figure 1.4. This is for (per Juran and Gryna, 24.33) total indicator readings for flatness or concentricity. Trace contaminants and impurities are not mentioned but another form of PRE-Control may work for them as well.

One-Sided Specification Limit: More Is Better

Juran and Gryna (24.34) provide an example in which a critical to quality specification limit, such as tensile strength, has a lower specification limit. The distribution may or may not be normal, but either way we do not care how much of the characteristic we get. We need some kind of upper limit to use as a reference, however, so we use the measurement from the best

---

* A threshold parameter means, for example, that we cannot get less than 0.2 parts per million of the undesirable characteristic in question.

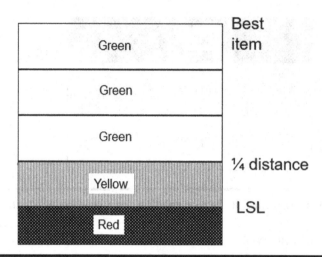

Figure 1.5 PRE-Control for lower specification limit, more is better

available item. The yellow zone consists of one-quarter of this distance from the lower limit, and the green zone the remaining three-quarters of the distance, as shown in Figure 1.5.

One-Sided Specification Limit: Less Is Better

Suppose instead that we have an upper specification limit, this time for trace contaminants or impurities. The PRE-Control scheme is simply the reverse of "more is better," as shown in Figure 1.6, noting that the best part (or sample) may have a measurement of greater than zero, or alternatively the lower detection limit of the instrument that is used for quality control.

I have not been able to find any literature that would suggest whether to use the figure for total indicator reading, in which the yellow zone is halfway between zero and the specification limit, or the one for "less is better," as shown here, for contaminants and impurities. This is apparently an unexplored application of PRE-Control. A key point is, however, that PRE-Control does enable qualification and monitoring of the process in question, which will allow the process owner to collect enough data to estimate the distribution's parameters and set up appropriate control charts.

Shop Floor Spreadsheet

Figure 1.7 shows a potential spreadsheet for nominal-is-best for use on the shop floor into which the production worker or inspector enters

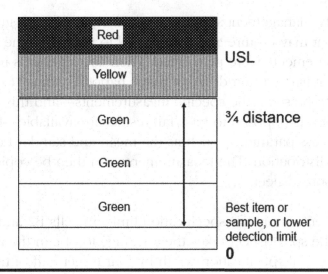

Figure 1.6   PRE-Control for upper specification limit, less is better

▲	A	B	C	D	E	F	G	H	I	J	K
1	PRE-Control, Nominal is Best										.
2	USL	0.520	User entry								
3	LSL	0.480	User entry								
4	Nominal	0.500	Calculated								
5	Delta (1/4 tolerance)	0.010	Calculated								
6											
7		Measurement					PRE-Control Zone				
8	Sample	1	2	3	4	5	R	Y	G	Y	R
9	Setup	0.502	0.493	0.511					XX	X	
10	Setup	0.494	0.497	0.507	0.498	0.502			XXXXX		
11	Monitor	0.502	0.485					X	X		
12	Monitor	0.495	0.503						XX		
13	Monitor	0.483	0.488					XX			
14	Adjust	0.495	0.498	0.507	0.493	0.505			XXXXX		
15	Monitor	0.495	0.515						X	X	
16	Monitor	0.502	0.487					X	X		
17											
18	Example: out of spec	0.495	0.473				X		X		
19	Example: out of spec	0.515	0.523							X	X
20											
21											

Figure 1.7   PRE-Control spreadsheet, nominal is best

measurements. The spreadsheet automatically classifies each measurement into the corresponding PRE-Control zone. While PRE-Control does not require the use of any charts whatsoever—remember that the monitoring step requires only that the production worker assess two parts—the ability to use a record of this nature covers two practical applications.

1. The quality management system, a customer, and/or a regulatory requirement may require a quality record to show that the necessary measurements were performed. There are situations in which the phrase "If it isn't written down, it didn't happen," comes to mind.
2. We want to preserve the specific measurements—and this assumes that real number as opposed to pass/fail results are available—for estimation of the process parameters such as the mean and standard deviation for a normal distribution. The measurements can then be copied directly from the spreadsheet.

The user need enter only the specification limits in cells B2 and B3 as shown here. The spreadsheet takes their average to obtain the nominal, and divides the total specification width by four to get half of the distance between the nominal and each specification limit. The production worker or inspector need enter only the measurements in the indicated columns and the spreadsheet does the rest. If we consider the first row of measurements:

■ Red (below LSL)=REPT("X",COUNTIF($B9:$F9,"<"&$B$3)) repeats the character X for every measurement in the row less than the lower specification limit in cell B3. COUNTIF($B9:$F9,"<"&$B$3) counts the entries that are less than the value in cell B3 while it ignores empty cells. The REPT function places an X in the cell for each measurement below the LSL.
  – We might in practice name the LSL, USL, nominal, and delta (1/4 tolerance width) values but the spreadsheet that accompanies the book has multiple examples, each with its own LSL and USL. If there is only one set, we might then use LSL instead of $B$3.
■ Yellow (low)=REPT("X",COUNTIFS($B9:$F9,">="&$B$3,$B9:$F9,"<"& $B$4-$B$5)) counts the measurements that are (1) greater than the lower specification limit in cell B3 and (2) less than the nominal minus half the distance between the nominal and the specification limit, or alternatively the nominal minus one-quarter of the total specification width. This is from cell B5.
■ Green=REPT("X",COUNTIFS($B9:$F9,">="&$B$4-B$5,$B9:$F9,"<="& $B$4+$B$5)) counts the measurements (1) greater than or equal to the nominal minus one-quarter of the specification width and (2) less than or equal to the nominal plus one-quarter of the specification width.
■ Yellow (high)=REPT("X",COUNTIFS($B9:$F9,">"&$B$4+$B$5,$B9: $F9,"<="&$B$2)) which counts the measurements (1) greater than the nominal plus one-quarter of the specification width and (2) less than or equal to the upper specification limit.

■ Red (above USL)=REPT("X",COUNTIF($B9:$F9,">"&$B$2)) which counts the measurements greater than the upper specification limit.

The first row shows that the third measurement was not in the green zone, so qualification of the process had to start over. The five green zone measurements in the next row qualify the process, whereupon monitoring with two-unit samples begins. The two yellow zone measurements indicate the need to adjust the process, whereupon another five-unit qualification is needed. Similar spreadsheets can be created for the total indicator reading, less is better, and more is better situations.

Spreadsheet for Less Is Better

Figure 1.8 shows how a spreadsheet for a quality characteristic where less is better can be deployed. Suppose the upper specification is 10 ppm of an undesirable trace contaminant, and the best sample (or the instrument's lower detection limit) is 1 ppm. The green zone then consists of 1 plus three-quarters of the distance to the USL, or 7.75.

◢	M	N	O	P	Q	R	S	T	U
1	PRE-Control, Less is Better								
2									
3	USL	10.000	User entry						
4	Best piece	1.000	User entry						
5	Delta (1/4 distance)	2.250							
6									
7		Measurement							
8	Sample	1	2	3	4	5	G	Y	R
9	Setup	3	4	3	2	8	XXXX	X	
10	Setup	5	3	1	7	6	XXXXX		
11	Sample	3	9				X	X	
12	Sample	8	2				X	X	
13	Sample	9	8					XX	
14	Adjust	3	5	4	2	6	XXXXX		
15	Sample	4	7				XX		
16	Sample	2	7.5				XX		
17									
18	Example: out of spec	6	10.5				X		X
19	Example: out of spec	10	11					X	X
20									
21									

Figure 1.8   PRE-Control, less is better

■ Cell S9 (green region)=REPT("X",COUNTIF($N9:$R9,"<="&$N$4+3*$N$5)) counts the entries that are less than or equal to the best piece (cell $N$4) plus three-quarters of the distance ($N$5) between the best piece and the upper specification limit.

■ Cell T9 (yellow region)=REPT("X",COUNTIFS($N9:$R9,">"&$N$4+3*$N$5, $N9:$R9,"<="&$N$3)) counts the entries that are greater than the upper limit of the green zone (best piece plus three-quarters of the distance to the USL) but less than or equal to the USL ($N$3).

■ Cell U9 (red)=REPT("X",COUNTIF($N9:$R9,">"&$N$3)) counts all the measurements that exceed the upper specification limit.

The other applications (total indicator reading and more is better) can be treated similarly to create both a visual control and also a quality record should one be necessary.

Summary: PRE-Control

PRE-Control is the simplest way to qualify and then monitor a process, and it requires no knowledge of the mean, standard deviation, or, in the case of nonnormal distributions, shape and scale parameters. It does not even require generation of a chart because the user can keep track of the periodic samples of two. PRE-Control may be used on its own, or to allow operation of a process until enough data accumulates with which to estimate the distribution parameters from which a traditional control chart can then be developed.

While PRE-Control is a form of process monitoring that does not require a control chart, a visual control can be added to create a quality record that includes the measurements of the parts along with their status. It requires only that the production worker enter the measurements; the spreadsheet can determine whether they fall into the green, yellow, or red zones. If, however, statistical control is required, then short-run SPC must be used.

Chapter 2

Introduction to Short-Run SPC

The previous chapter showed how PRE-Control can be used to monitor a process, and with only prior knowledge of its specification limits. This chapter will show how to set up simple Shewhart control charts that use statistical parameters (a nominal and a standard deviation) to define their control limits. AIAG (2016, 9–10) offers two very simple approaches:

1. Rely on a surrogate process for which historical information is available.
2. Rely on target parameters, i.e. what the mean and standard deviation ought to be. The target mean is simply the nominal, which is halfway between the specification limits. The process standard deviation σ_{process} can also be calculated from the process performance index P_{p} as shown in Equation 2.1.

$$P_{\text{p}} = \frac{\text{USL} - \text{LSL}}{6\sigma_{\text{process}}} \rightarrow \sigma_{\text{process}} = \frac{USL - LSL}{6P_{\text{p}}} \tag{2.1}$$

If we again paraphrase the conversation between Big Jule and Nathan Detroit in Guys and Dolls, how can we determine the process standard deviation when we don't know the process performance index, which is itself a function of the process standard deviation? Big Jule claimed to know where the spots on the dice had been, and we know with far more justification what P_{p} ought to be. A process is generally not considered capable unless P_{p}

DOI: 10.4324/9781003281061-2

is 4/3 or greater, and customers may require 5/3 or even 6/3 (Six Sigma). If we assume that the requirement is 4/3, then:

$$\frac{4}{3} = \frac{\text{USL} - \text{LSL}}{6\sigma_{\text{process}}} \rightarrow \sigma_{\text{process}} = \frac{\text{USL} - \text{LSL}}{8}$$

It's that simple. A 4-sigma process has four standard deviations between its nominal and each specification limit, so there are eight standard deviations or "sigmas" between the specification limits. If, for example, the specification limits are 64 and 66, respectively, then we assume the standard deviation to be 0.25. This allows us to develop control charts for sample averages and sample ranges (R charts) or sample standard deviations (s charts).

Limitation to Normally Distributed Processes

This also shows, however, that this approach is not usable for non-normal distributions except for the exponential distribution, which has only one parameter. The issue is not that we cannot calculate process performance indices and control limits for non-normal processes; off the shelf methods are available for both. Shewhart-equivalent control limits can be set at the 0.00135 and 0.99865 quantiles of the distribution, and the center line placed at the median (0.50 quantile) rather than the mean. The obstacle is that we have two unknowns, the distribution's shape and scale parameters, and only one equation for the required nonconforming fraction as a function of the two unknowns. A second equation can be provided if the target mean also is given, but these distributions often have one-sided specification limits which make the target either as little as possible or as much as possible.

The mean and standard deviation of the gamma distribution are, for example, both functions of its shape and scale parameters, and different combinations of the two can meet the requirement that the process generates no more than 31.7 nonconformances per million opportunities, the same quantity that a normally distributed four sigma process delivers at each specification limit.

$$f(x) = \frac{\gamma^{\alpha}}{\Gamma(\alpha)} x^{\alpha-1} \exp(-\gamma x) = \frac{\left(\frac{1}{\beta}\right)^{\alpha}}{\Gamma(\alpha)} x^{\alpha-1} \exp\left(-\frac{1}{\beta}x\right) \text{ Gamma distribution}$$

- Alpha (sometimes known as k) is the shape parameter. The gamma distribution is in fact the probability density function for the sum of α (or k) random variables from an exponential distribution with the scale parameter cited below.
- Gamma is the scale parameter, as used by StatGraphics.
- Beta is the scale parameter, as used by Excel's gamma distribution functions.

Suppose, for example, that a process follows a gamma distribution, and its upper specification limit is 5 (parts per million of a trace contaminant, for example). Also assume that one-sided 4-sigma process performance (*Ppk*) of 4 is required. Such a process has 31.7 nonconformances per million at each specification limit. This requirement can be fulfilled with a gamma distribution whose shape parameter is 2 and scale parameter is 2.6, as shown in Figure 2.1 from StatGraphics.

Figure 2.2 shows, however, that a shape parameter of 3 and a scale parameter of about 3.05 also will work.

This shows that an almost limitless combination of shape and scale parameters will meet the process capability requirement, and there is no way to set up a control chart without prior knowledge of one of them. That

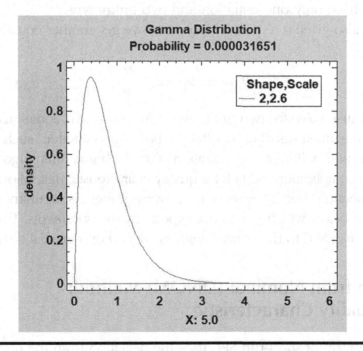

Figure 2.1   Nonconforming fraction above USL = 5, shape = 2, and scale = 2.6

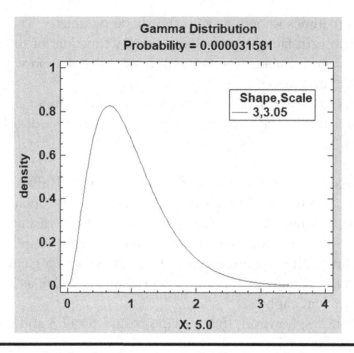

Figure 2.2 Nonconforming fraction, gamma distribution with shape 3, scale 3.05

is, if F is the cumulative gamma distribution, then F(USL=5 ppm given α and γ)=31.7 per million has an infinite number of solutions for α and γ because we have only one equation and two unknowns.

If we are also given a required mean, then we get another equation, namely:

$$\mu_{gamma\_distribution} = \frac{\alpha}{\gamma}$$

We can of course solve for two unknowns if we have two equations, but applications of the gamma distribution often involve "less is better," such as trace contaminants and pollutant concentrations for which it is often a good model.

The same complication exists for a quality characteristic that follows the Weibull distribution. Matters become even worse if there is an unknown threshold parameter below which we do not expect any measurements. This limits traditional short-run SPC to the normal, lognormal, and exponential distributions.

Deviation from Nominal (DNOM) Method: Single Quality Characteristic

The simplest form of short-run SPC uses the deviation from the nominal method (AIAG, 2016, Chapter 1). Applications consist of processes with *one* set of specification limits and *one* process standard deviation. This chapter will focus only on the latter. Subsequent chapters will address the applications that involve multiple specification limits, multiple product features, and/or part or feature-specific standard deviations. The process standard deviation comes from either (1) surrogate or similar processes or (2) the specification limits and required process performance index, as discussed previously. The deviation from nominal (dnom) is then simply the individual measurement X or the sample average x-bar minus the nominal, and the center line is then zero.

$$\text{dnom} = X - \mu_{\text{target}} \text{ or dnom} = \overline{x} - \mu_{\text{target}}$$

The AIAG reference uses the calculated average range, where d_2 is the control chart factor for a sample of size n, to find the control limits for the charts for sample average and sample range.

$$\overline{R} = d_2 \sigma_{\text{process}}$$

If only individual measurements are available, the average moving range is used, where

$$\overline{MR} = d_2 \sigma_{\text{process}} = 1.128 \sigma_{\text{process}} = \frac{2}{\sqrt{\pi}} \sigma_{\text{process}}$$

Then the control limits of the chart for individuals or sample averages are:

$$0 \pm E_2 \overline{MR} = 0 \pm 3 \frac{\sqrt{\pi}}{2} \overline{MR} \quad \text{for } n = 1 \quad 0 \pm A_2 \overline{R} \quad \text{for } n \geq 2$$

This can, however, be simplified enormously to avoid the need to look up control chart factors, even though Excel can do this automatically if a table is provided.

$$0 \pm 3 \frac{\sqrt{\pi}}{2} \overline{MR} = 0 \pm 3 \frac{\sqrt{\pi}}{2} \frac{2}{\sqrt{\pi}} \sigma_{\text{process}} = 0 \pm 3 \sigma_{\text{process}}$$

$$0 \pm A_2 \overline{R} = 0 \pm A_2 d_2 \sigma_{\text{process}} = 0 \pm \frac{3}{\sqrt{n}} \sigma_{\text{process}}$$

The simplest form results, however, through the use of the *standard normal deviate* of the deviation from nominal. This allows, in fact, the use of non-uniform sample sizes. The control limits of the standard normal deviate chart are simply ±3 regardless of the sample size *n*. Another advantage is that we can dispense with the control chart factors entirely, as shown in Equation 2.2.

$$z = \frac{dnom}{\dfrac{\sigma_{process}}{\sqrt{n}}} = \frac{\bar{x} - nominal}{\dfrac{\sigma_{process}}{\sqrt{n}}} \tag{2.2}$$

Chart for Process Variation

We have seen so far that it is extremely easy to handle the X chart for individuals or x-bar chart for sample averages through the use of the standard normal deviate. AIAG (2016, Chapter 1) shows how to develop range charts for samples of two and larger where D_3 and D_4 are the control chart factors for samples of size *n*.

$$\text{LCL} = D_3 \bar{R} \quad \text{and} \quad \text{UCL} = D_4 \bar{R}$$

There is, however, no need to calculate the average range because, where D_1 and D_2 are again control chart factors that depend on the sample size,

$$\text{LCL} = D_1 \sigma_{process} \quad \text{and} \quad \text{UCL} = D_2 \sigma_{process}$$

Suppose, however, we have only individual measurements. The AIAG reference uses the moving range chart, but there are questions as to just how useful this is. The average moving range itself is very useful for estimation of the short-term process variation (the foundation of process capability rather than performance indices) but ASTM (1990, 97) says, "All the information in the chart for moving ranges is contained, somewhat less explicitly, in the chart for individuals." AT&T (1985, 22) adds, "Do not plot the moving ranges calculated in step (2)." This suggests that the moving range chart is not particularly useful and need not be plotted at all.

Example: Chart for Individuals

A product's specification limits are 99 and 101 respectively, and the customer requires a process performance index of 5/3, i.e. a 5-sigma process. The process mean is actually 99.7 and its standard deviation is really 1/6, i.e. a

6-sigma process, but the process owner has no way to know this. Start with the assumption that the process meets the 5-sigma requirement, and that it is centered on its nominal.

$$\sigma_{\text{process}} = \frac{\text{USL} - \text{LSL}}{6P_p} = \frac{101 - 99}{6\frac{5}{3}} = \frac{1}{5}$$

The spreadsheet that accompanies the book includes 50 simulated data whose mean is 99.7 and standard deviation is 1/6 in the sheet DNOM_ Individuals. StatGraphics addresses this application with the "control to standard" option for the control chart for individuals, as shown in Figure 2.3.

Figure 2.4 displays an out of control signal below the lower control limit, which tells us that the process is not centered on the nominal.

Deployment to the Shop Floor

The StatGraphics control chart was obtained from entry of all 50 simulated data, but what we need on the shop floor is a spreadsheet into which the inspector or production worker can enter measurements one at a time and get immediate feedback on whether an out of control signal has occurred.

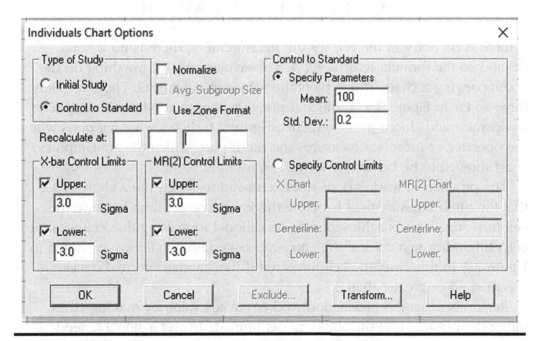

Figure 2.3 Control to standard in StatGraphics

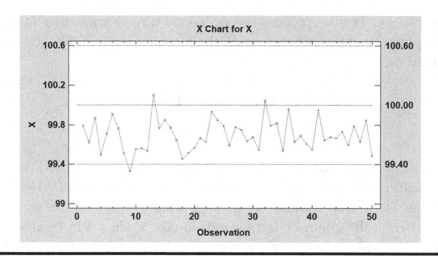

Figure 2.4   StatGraphics control chart for individuals

Figure 2.5 shows one way to set up an Excel spreadsheet to accept up to 50 measurements on a one-at-a-time basis, which is what is actually needed on the shop floor. All that the production worker or inspector needs to do is enter the measurement and also possibly identifying information in the description field, and the spreadsheet will do the rest.

The formula in cell G9, which can be copied to those below it, is:

$$= IF(ISBLANK(F9), NA(), (F9\text{-}F\$5)/F\$4)$$

If there is no entry in the cell for the measurement, there is no z statistic, and so the formula returns #N/A. This avoids plotting anything on the accompanying z chart, which plots the values in column G. There are only three so far in Figure 2.5. Additional points will appear when the inspector or operator adds more measurements. Figure 2.6 shows what happens when the operator or inspector measures the ninth part, which the StatGraphics chart showed to be below the lower control limit.

The production worker does not even need to watch the z chart to see that the ninth measurement is below the lower control limit because the cell turns red. This is achieved with conditional formatting that changes the cell's fill and/or text color if the entry meets certain conditions as shown in Figure 2.7. In this case, the cell turns red if the standard normal deviate is less than −3 or greater than 3.

Remember that the production worker creates value by making parts rather than charts, and the ability to get immediate and actionable feedback by entering just the measurement, along with any required identification (such as the date and time of the measurement) supports this practical

E	F	G	H	I	J	K	L
USL	101	User entry					
LSL	99	User entry					
Required Pp	1.666667	User entry					
sigma_process	0.2	calculated					
nominal	100	calculated					
z Chart							
Description	X	z					Z
1	99.790	-1.05					
2	99.620	-1.90	4.00				
3	99.868	-0.66					
		#N/A	3.00				
		#N/A	2.00				
		#N/A					
		#N/A	1.00				
		#N/A	0.00				
		#N/A					
		#N/A	-1.00				
		#N/A					
		#N/A	-2.00				
		#N/A					
		#N/A	-3.00				
		#N/A					
		#N/A					

Figure 2.5   Deployment of the z chart in Excel

consideration. This approach means the worker does not even have to interpret a control chart, although one can be added easily. If the cell turns red,
the problem is immediately obvious; the process mean has probably shifted.
(The qualifier "probably" is present because of the false alarm risk associ-
ated with control charts.) This tells the process owner to adjust the process
before it can produce any nonconforming work.

Charts for Samples

The next step is to address applications in which samples of two or greater
are available. We already have the standard normal deviate:

$$z = \frac{\overline{x} - \text{nominal}}{\dfrac{\sigma_{\text{process}}}{\sqrt{n}}}$$

	E	F	G	H	I	J	K	L
USL		101	User entry					
LSL		99	User entry					
Required Pp		1.666667	User entry					
sigma_process		0.2	calculated					
nominal		100	calculated					
z Chart								
Description		X	z					
	1	99.790	-1.05					
	2	99.620	-1.90					
	3	99.868	-0.66					
	4	99.4936	-2.53					
	5	99.71061	-1.45					
	6	99.90934	-0.45					
	7	99.76202	-1.19					
	8	99.51421	-2.43					
	9	99.33076	-3.35					
			#N/A					
			#N/A					
			#N/A					
			#N/A					
			#N/A					
			#N/A					

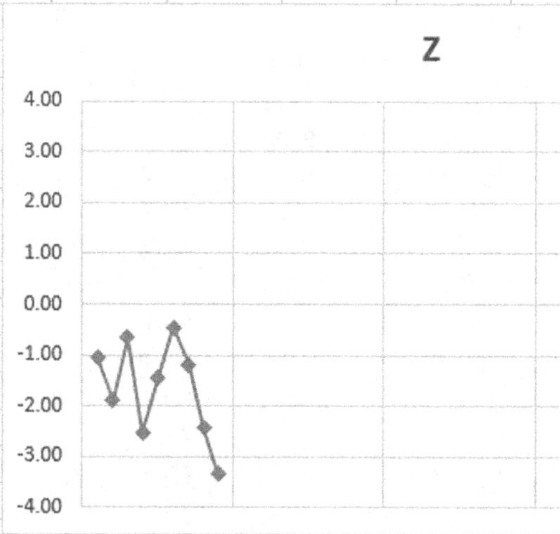

Figure 2.6   Detection of out of control signal

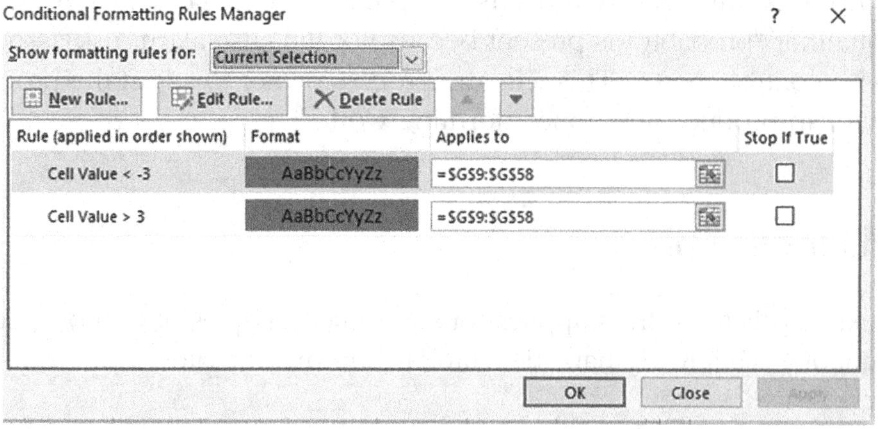

Figure 2.7   Conditional formatting of the z chart

AIAG (2016, 11) shows how to create a standardized range chart whose control limits are simply the D_3 and D_4 control chart factors, respectively, for a sample of size n, and the standardized sample range is:

$$R_{\text{standardized}} = \frac{R}{\bar{R}} \quad \text{where} \quad \bar{R} = d_2 \sigma_{\text{process}}$$

This can be handled in Excel by having the VLOOKUP or HLOOKUP function select the right control chart factors from a table, and different sample sizes can even be addressed. Suppose, for example, that a work cell's bill of materials, or kanban, calls for two units of A every odd hour and three units of A every even hour. The specification limits are still 99 and 101 respectively and the required process performance index is 5/3. The process is now centered on the nominal but the process is in fact not capable because its standard deviation is 1/3, i.e. a 3-sigma process. Figure 2.8 shows some of the simulated data.

X1	X2	X3
99.832	100.010	
99.649	100.047	99.773
99.574	99.693	
100.150	100.134	100.539
100.402	100.043	
99.858	99.919	99.607
99.814	99.575	
100.173	100.009	100.064
99.978	99.728	
100.025	99.816	99.759
99.467	99.617	
100.339	99.407	99.873
99.807	100.224	
100.353	99.423	99.258

Figure 2.8   Data for samples of different sizes

StatGraphics will generate charts with sample-specific control limits given Control to Standard, mean=100, standard deviation=0.2, and "Avg. Subgroup Size" unchecked. Figure 2.9 shows what these look like. The range chart has no lower control limit for sample sizes of two or three, and the lower line in the range chart is the center line for the sample size in question. The center line itself is a function of the sample size.

The points above the range chart's upper control limit, and we would react to the 12th, show that the process standard deviation is greater than the required 0.2. The wide swings in the *x*-bar chart show that the control

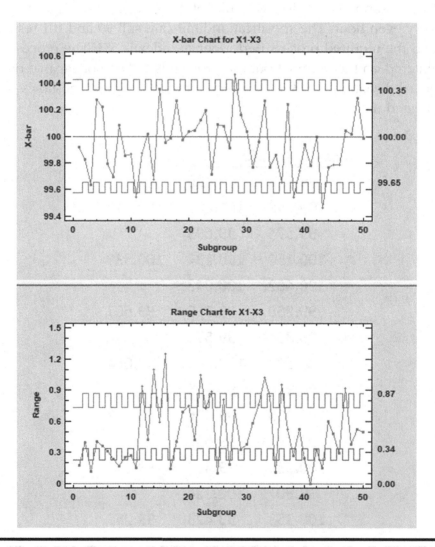

Figure 2.9   Control charts for variable sample sizes

limits are too tight. They assume that the process standard deviation is 0.2, and it is really 0.333. (When the chart for process variation, such as the sample range or sample standard deviation chart, indicates that the variation has increased, it is standard practice to assume that the control limits for the *x*-bar chart are too tight. That is, the R or s chart is interpreted before the *x*-bar chart.)

The *x*-bar chart can be deployed in Excel as shown in Figure 2.10, with the same conditional formatting that was discussed earlier. The cell will turn red if the standard normal deviate z is less than −3 or greater than 3, and z is now calculated as follows for cell I9. =IF(ISBLANK(F9),NA(),(AVERAGE(F 9:H9)-F$5)/(F$4/SQRT(COUNT(F9:H9)))) puts #N/A in the cell if there is no measurement for X1. If there is, divide the difference between the sample average and the nominal (F5) by the process standard deviation (F4) divided by the square root of the sample size as returned by the COUNT function. This counts the number of cells in the range F9 through H9 for which there are entries.

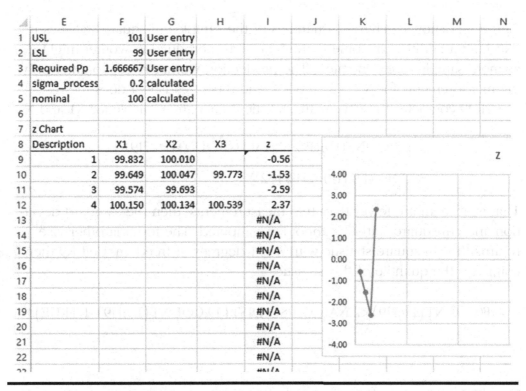

	E	F	G	H	I	J	K	L	M	N
1	USL	101	User entry							
2	LSL	99	User entry							
3	Required Pp	1.666667	User entry							
4	sigma_process	0.2	calculated							
5	nominal	100	calculated							
6										
7	z Chart									
8	Description	X1	X2	X3	z					
9	1	99.832	100.010		-0.56					z
10	2	99.649	100.047	99.773	-1.53		4.00			
11	3	99.574	99.693		-2.59		3.00			
12	4	100.150	100.134	100.539	2.37					
13					#N/A		2.00			
14					#N/A		1.00			
15					#N/A					
16					#N/A		0.00			
17					#N/A		-1.00			
18					#N/A					
19					#N/A		-2.00			
20					#N/A		-3.00			
21					#N/A					
22					#N/A		-4.00			
23					#N/A					

Figure 2.10   Standard normal deviates, variable sample size

Probabilistic Control Limits for Variation

The next question is whether we should set up variable control limits for the sample ranges by using Excel's LOOKUP functions to consult tables of control chart factors. We can, however, simplify matters enormously by using *probabilistic control limits*, which is in fact how StatGraphics handles control charts for non-normal distributions. AIAG (2005, 115-116) adds that we can, for non-normal distributions, set the control limits at the distribution's 0.135 and 99.865 percentiles to match those of the 3-sigma limits of a Shewhart chart for normally distributed data. We can in fact select any false alarm risk we want. This is deployable to the sample standard deviation instead of its range through the use of the chi-square statistic.* The chi-square test statistic has $n - 1$ degrees of freedom for a sample of n items, where $(n - 1)$ times the square of the sample standard deviation is divided by the square of the target or hypothetical standard deviation as shown in Equation 2.3.

$$\chi^2_{n-1} = \frac{(n-1)s^2}{\sigma_0^2} = \frac{(n-1)s^2}{\sigma^2_{\text{process}}} \tag{2.3}$$

Figure 2.11 shows a chart of quantiles of the sample chi-square statistics in Excel, after addition of some more data, as the production worker might do as more samples are generated. The production worker needs to enter only the measurements for X1, X2, and X3, and the spreadsheet does the rest.

Cell J9 computes the chi-square statistic if data are available for X1 and X2.

=IF(COUNT(F9:H9)<2,NA(),((COUNT(F9:H9)-1)

*STDEV(F9:H9)^2/F$4^2))

That is, if there are fewer than two measurements, then the standard deviation and chi-square statistic cannot be computed. The formula otherwise returns the chi-square statistic with $n - 1$ degrees of freedom. Cell K9 then computes the quantile of this statistic.

$$=IF(COUNT(F9:H9)<2,NA(),CHISQ.DIST(J9,COUNT(F9:H9)-1,TRUE))$$

---

* It is also possible to calculate specific quantiles for sample ranges, but this requires programming in, for example, Visual Basic for Applications. Excel has, on the other hand, a built-in function for quantiles of the chi-square distribution.

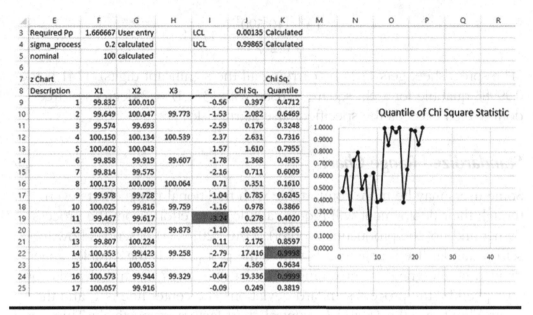

	E	F	G	H	I	J	K
3	Required Pp	1.666667	User entry		LCL	0.00135	Calculated
4	sigma_process	0.2	calculated		UCL	0.99865	Calculated
5	nominal	100	calculated				
6							
7	z Chart						Chi Sq.
8	Description	X1	X2	X3	z	Chi Sq.	Quantile
9	1	99.832	100.010		-0.56	0.397	0.4712
10	2	99.649	100.047	99.773	-1.53	2.082	0.6469
11	3	99.574	99.693		-2.59	0.176	0.3248
12	4	100.150	100.134	100.539	2.37	2.631	0.7316
13	5	100.402	100.043		1.57	1.610	0.7955
14	6	99.858	99.919	99.607	-1.78	1.368	0.4955
15	7	99.814	99.575		-2.16	0.711	0.6009
16	8	100.173	100.009	100.064	0.71	0.351	0.1610
17	9	99.978	99.728		-1.04	0.785	0.6245
18	10	100.025	99.816	99.759	-1.16	0.978	0.3866
19	11	99.467	99.617		-3.24	0.278	0.4020
20	12	100.339	99.407	99.873	-1.10	10.855	0.9956
21	13	99.807	100.224		0.11	2.175	0.8597
22	14	100.353	99.423	99.258	-2.79	17.416	0.9998
23	15	100.644	100.053		2.47	4.369	0.9634
24	16	100.573	99.944	99.329	-0.44	19.336	0.9999
25	17	100.057	99.916		-0.09	0.249	0.3819

Figure 2.11   Control chart for process variation

If there were fewer than two measurements, then this cannot be performed; the function otherwise returns the *cumulative probability* of the chi-square statistic in question. The cell's conditional formatting will turn the cell red if this quantile exceeds the upper control limit, which is 1 minus the user-specified false alarm risk (alpha risk) divided by 2. It will turn it green if it falls below the lower control limit, which is the user-specified false alarm risk divided by 2. A point below the LCL for a range or sample standard deviation chart could mean that a process improvement whose purpose was to reduce variation worked, although this is otherwise likely to be due to random or common cause variation. That is, if the one-sided risk is 0.00135, then we expect 1.35 points out of every 1000 to fall below the LCL. The accompanying control chart for the quantiles has a center line of 0.50, and we expect the points to scatter randomly around this line.

Consider, for example, the 14th point, which resulted in an out of control signal on the chart for process variation. The average of 100.353, 99.423, and 99.258 is 99.678, and their standard deviation is 0.590. Then, given a nominal of 100 and presumed process standard deviation of 0.2,

$$z = \frac{99.678 - 100}{\dfrac{0.2}{\sqrt{3}}} = -2.79$$

$$\chi^2 = \frac{(3-1)0.590^2}{0.2^2} = 17.41$$

(The spreadsheet gets 17.24 by carrying all the significant digits). 17.41 is the 0.99983 quantile for a chi-square statistic with $(3 - 1) = 2$ degrees of freedom, which exceeds the specified false alarm risk.

Standardized Range Chart

Some users may be more comfortable with traditional range charts even though their deployment will require a little more effort (from the process owner and not the production worker). AIAG (2016, 11) describes how to set up standardized range charts where the plot point is the sample range divided by the grand average range. The latter is the d_2 control chart factor multiplied by the process standard deviation. The center line is 1, and the control limits are the D_3 and D_4 control chart factors, respectively. This can be achieved as shown in Figure 2.12.

Cell J9 contains the standardized range, =IF(ISBLANK(F9),NA(),(MAX(F9: H9)-MIN(F9:H9))/(VLOOKUP(COUNT(F9:H9),I$2:L$5,2)*F$4)). If there are no

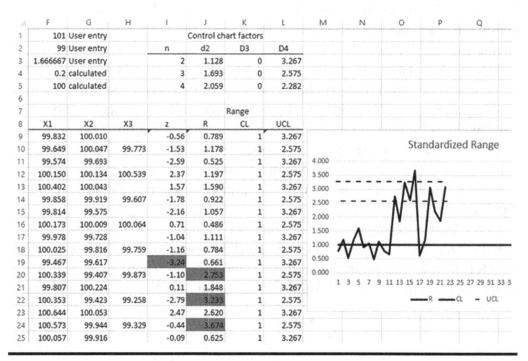

Figure 2.12 **Standardized range chart**

data, as reflected by no entry in the cell for the first measurement (F9), the function returns #N/A. Otherwise it computes the range as the maximum of cells F9 through H9 minus their minimum; the MAX and MIN functions ignore blank cells. The result is divided by the product of the control chart factor d_2 as obtained from the VLOOKUP function on the table of control chart factors and the process standard deviation in cell F4.

$$\frac{100.01 - 99.832}{1.128 \times 0.20} = 0.789$$

The center line is simply 1 and the upper control limit is=IF(ISBLANK(F9),N A(),(VLOOKUP(COUNT(F9:H9),I\$3:L\$5,4))), where VLOOKUP looks for the number of measurements (two or three) as determined by the COUNT function, and cross-references the result for the fourth column (D_4 factors) in the table of control chart factors. This allows generation of the standardized range chart. A short horizontal line was selected as the marker for the upper control limit.

The conditional formatting rule for Column J was created by placing the following (Figure 2.13) in cell J9, and then copying the format down the rest of the column. The selected fill color is red which appears as gray in the figure.

Remember, however, that the purpose of SPC is to facilitate the production of parts as opposed to charts. The visual control of a change in cell color when a statistic exceeds its control limits should be sufficiently adequate by itself to alert the production worker that whatever reaction plan (part of the control plan) is in place should be used to correct the situation. The selection of any accompanying charts should accord with the organization's experience with control charts including production worker familiarity

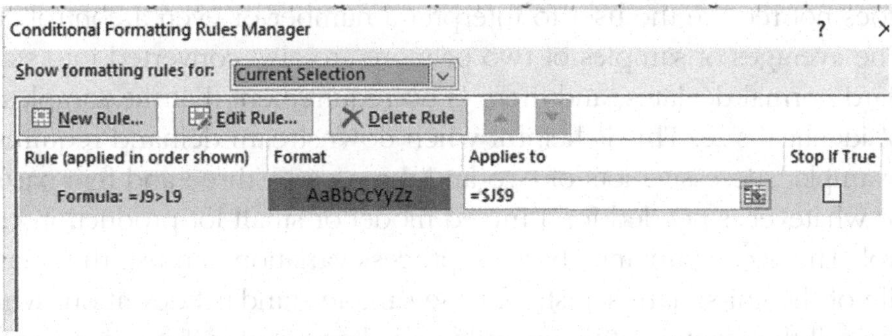

Figure 2.13   Conditional formatting for standardized ranges

with their formats. Also note that, regardless of whether we use the chi-square statistic with probabilistic limits, or the standardized range chart with traditional limits from control chart factors, *the user needs to enter only the measurements for each sample to get immediate and actionable feedback.*

Summary: Short-Run SPC, Single Quality Characteristic

This chapter has addressed the simplest applications for which one quality characteristic has a single set of specification limits, and therefore a single target or nominal, along with a single standard deviation $\sigma_{process}$. The latter can come from surrogate processes, or it can be calculated from the specification width on the basis of a required process performance index. The latter approach is limited to processes that follow the normal (bell curve) distribution and also the exponential distribution, as the latter has only one parameter. The lognormal distribution also works because the natural logs of the measurements follow the normal distribution.

The chapter's vital takeaway is that the resulting short-run SPC chart can be deployed to the shop floor on a spreadsheet, i.e. specialized software that might not even be designed for routine process control is not required. All that the worker or inspector needs to do is enter the measurements along with any required descriptive information (such as date, time, and part identification) for a quality record. The spreadsheet then does the following:

1. Individual measurements are converted to their standard normal deviates (z statistics) for which the X chart's center line is 0 and the control limits ±3. A control chart can be plotted but the cell that contains the z statistic can be programmed, via conditional formatting, to turn red for results outside the control limits. This is an ideal *visual control* that does not require the user to interpret a number or even a control chart.
2. The averages of samples of two or more are also converted into standard normal deviates, and there is no requirement that the samples be of identical size. This is helpful when downstream demand requires, for example, successive lots of two and three parts, three and five parts, or whatever is needed for a mixed model or small lot production control. The accompanying chart for process variation can use the quantile of the chi-square statistic for the sample standard deviation, while the cell that contains this quantile will change color if it is outside the

user-specified limits for false alarms. The quantiles will, if the process standard deviation is indeed the same as the target, follow the uniform distribution; this format may be unfamiliar to people who are familiar with traditional Shewhart charts. A standardized range or standard deviation chart can be used instead if this is a problem.

The next chapter will introduce the cumulative sum (CUSUM) and exponentially weighted moving average (EWMA) charts whose purpose is to detect small process shifts and is ideally suited to the monitoring of startup processes.

Chapter 3

Cumulative Sum (CUSUM) Chart

The CUSUM chart is useful for detecting small changes from a nominal or target, which makes it a useful variation on the DNOM chart. This chapter will also cover the exponentially weighted moving average (EWMA) chart which is similar to CUSUM but is designed so older measurements have less influence on the test statistic.

Engineering Process Control and CUSUM

Box and Luceño (1997, Chapter 6) show there is a relationship between CUSUM, EWMA, and the kind of automatic process control or engineering process control that is used in the chemical process industries. The key take-away is that CUSUM, EWMA, and integral control (which will be discussed below) all measure in some way *accumulated deviations from the target or nominal*, in contrast to assessment of only the current measurement or sample statistic. EWMA differs from CUSUM because it gives less weight to older measurements.

Consider simple proportional control, and let W_t be a process variable (or compensatory variable) at time t whose adjustment w_t influences the process output x. W_t can be, for example, a metal deposition rate as cited in the reference, a tool speed, or a spin coating speed. The proportional control equation for w_t at time t is

$$gw_t = -Ge_t$$

DOI: 10.4324/9781003281061-3

where e_t is the error, difference between the process output and the target or nominal, or deviation from nominal (dnom). G is the *damping factor* whose purpose is to avoid overadjustment. g is the *process gain*, or change in process output in response to an incremental change in process variable W. If the variable of interest is x, then:

$$\frac{dx}{dW} = g$$

Assume for convenience that $g=1$, i.e. we can adjust the process output directly, and that G also is 1. We then have adjustment $w_t = -e_t$ which is Rule 2 for W. Edwards Deming's funnel experiment. If, for example, the process output is 1 unit above the nominal, we adjust the process' point of aim by -1. Box and Luceno (1997, 133) refer to this as *full adjustment* and also cite Deming in the context of tampering, overadjustment, or tinkering.

Full adjustment is perhaps the simplest form of proportional control and, as shown by the funnel experiment, it increases rather than decreases the variation in the system. The explicit purpose of the damping factor G is to avoid overadjustment but, in the absence of anything but common cause variation, proportional control will nonetheless increase overall variation. Suppose, for example, that a process follows an integrated moving average (IMA) time series model in which Z_t, the disturbance at time t, is given by the following model where a is normally distributed white noise, and theta is between 0 and 1:

$$Z_t = Z_{t-1} + a_t - \theta a_{t-1}$$

If theta is 1, then there is no time series behavior whatsoever because

$$Z_0 = a_0 \text{ and then } Z_1 = a_0 + a_1 - a_0 = a_1,$$

which makes the disturbances independent of one another and also makes their variance equal to that of the white noise. Equation 6.12 of the reference shows that the ratio of the overall variance of the process to that of the white noise is then:

$$\frac{\sigma_e^2}{\sigma_a^2} = 1 + \frac{\left(G - (1-\theta)\right)^2}{G(2-G)}$$

If theta equals 1, then only white noise is present in the system, which makes any form of adjustment undesirable. Rule 2 of Deming's funnel experiment, which uses $G=1$, results in a variance ratio of 2. That is, the long-term variance will be twice the white noise variance. The same equation shows, however, that if theta is less than 1 and there is IMA behavior, the long-term variance can be removed entirely if G can be set exactly equal to $(1-\theta)$. The reference (pp. 149–151) also describes how to suppress variation in a first-order autoregressive model. The reference (pp. 131–133) describes how a five-fold reduction in mean square error (MSE, the average of the squared deviations from the target) was achieved for a semiconductor metal deposition process through a manual proportional control approach.

The details of this kind of adjustment scheme are beyond the scope of this book but the valuable takeaway is that it is indeed possible to adjust systematic forms of time series-related variation out of a process if the time series behavior can be quantified.

Integral Control

Proportional control can be combined with integral and derivative control to obtain proportional-integral-derivative (PID) control. Derivative control, which bases control action on the rate of change in the error, is not of further relevance here but integral control is. Integral control uses the following model where g is again the gain, k_0 an independent constant, and k_I a constant for integral control.

$$gw_t = k_0 + k_I \int e_t dt$$

Recall that e_t is the difference between the process output and the target at time t. While proportional control acts on the current error, or deviation from nominal, integral control acts on the integral of the errors over a period of time. CUSUM is very similar except for the fact that it uses discrete summations instead of integration of continuous data.

CUSUM

The following discussion is only for reference because practical deployment on a spreadsheet is problematic; we will use instead a tabular version that is much more spreadsheet friendly. The quantity plotted for the *i*th measurement is, per Montgomery (1991, 280–285), and the formula also works for individual measurements, as follows.

$$S_i = \sum_{j=1}^{i} \left(\overline{x}_j - \mu_{\text{target}} \right) = \sum_{j=1}^{i} \text{dnom}_j = \sum_{j=1}^{i} e_j$$

The last version of this equation, which uses the sums of the errors, relates the cumulative sum to the equation for integral control.

The key takeaway here is that the cumulative sum S_i *accumulates* the deviations from nominal, which increases the chart's ability to detect small deviations from nominal. The chart uses a V-shaped (rotated counterclockwise) mask, or set of control limits, to detect relatively small process shifts which makes it ideal for process startups where the idea is to control to nominal. The V-mask can be customized for the user-specified (1) *two-sided* false alarm risk alpha (α) of concluding that the process mean has changed, and (2) risk beta (β) of not detecting a specified shift of delta (δ) standard deviations. The deliverables are (1) the distance d between the vertex of the V-mask and the most recent point and (2) theta (θ), which is half the angle of the V-mask at its vertex. We might, however, dispense with theta because the real quantity of interest is the slope of the lines of the V-mask.

$$d = \left(\frac{2}{\delta^2} \right) \ln \left(\frac{1-\beta}{\frac{\alpha}{2}} \right) \text{ or } -\left(\frac{2}{\delta^2} \right) \ln \left(\frac{\alpha}{2} \right) \quad \text{where } \beta \text{ is very small}$$

Note the change of sign from + to − when $\alpha/2$ moves to the numerator.

$$\theta = \tan^{-1} \left(\frac{\delta\sigma}{2A} \right)$$

where A is between 1 and 2 times the standard deviation of the sample or individual measurement, with the recommendation being 2. A is a *scale factor* that relates the vertical and horizontal scale units (Montgomery, 1991, 285). Remember that the standard deviation of a sample of n measurements is simply that of an individual divided by the square root of n.

Montgomery (280–285) elaborates, and this summary will be useful as we move forward,

■ A is the scale factor, and a multiple of the process standard deviation. Its units are, therefore, like those of the standard deviation, the same as that of the quantity under consideration.

■ *d* is the lead distance between the most recent axis point (e.g. the axis point of the 10th sample if we have 10 samples) and the vertex of the V-mask.

■ Δ (Delta) is the process shift we want to detect. Its units of measurement are the same as the quality characteristic.

■ δ (delta) is Δ divided by the sample standard deviation and is therefore unitless. If, for example, the standard deviation of an individual is 0.4 mils, and we use a sample of 4, the standard deviation of the sample average is 0.2 mils (0.4 divided by the square root of 4). If we want to detect a shift of Δ=0.1 mil in the process mean, then δ = (0.1 mil)/(0.2 mil)=0.5. Alternatively, Δ=δ ×σ $_{sample}$.

■ *H* is the decision interval, which is the vertical distance between the most recent axis point and either line of the V-mask. *H*=*A*×*d*×tan(θ) and therefore:

$$\theta = \tan^{-1}\left(\frac{\delta\sigma}{2A}\right) \text{ so } \tan(\theta) = \frac{\delta\sigma}{2A} \text{ and then } H = \frac{d\delta\sigma}{2}$$

Montgomery (1991, 286–287) presents an example in which the objective is to detect a 1.5-sigma process shift when sigma is 1 and *d*, as calculated previously, is 4.71. 1.5×4.71/2=3.53, which is the result when one carries all the significant figures in the example. *This leaves us with the following convenient set of Equations (3.1) for the key deliverables H and d, and we don't have to worry about theta.* Note that alpha is the two-sided false alarm risk.

$$d = \left(\frac{2}{\delta^2}\right)\ln\left(\frac{1-\beta}{\frac{\alpha}{2}}\right) \text{ or } -\left(\frac{2}{\delta^2}\right)\ln\left(\frac{\alpha}{2}\right) \quad \text{when } \beta \text{ is very small} \quad (3.1a)$$

$$H = \frac{d\delta\sigma}{2} \tag{3.1b}$$

■ *h* is the decision interval expressed in multiples of the sample standard deviation, so *H*=*h*×σ $_{sample}$.

■ *K* is the absolute value of the slope of either side of the V-mask. If we recall that *H* is the vertical distance between the axis point of the most recent sample, and *d* the lead distance between that point and the vertex of the V-mask, then the absolute value of *K* is simply *H/d*.

■ *k* is the slope in terms of multiples of the sample standard deviation, i.e. *K*=*k*×σ $_{sample}$.

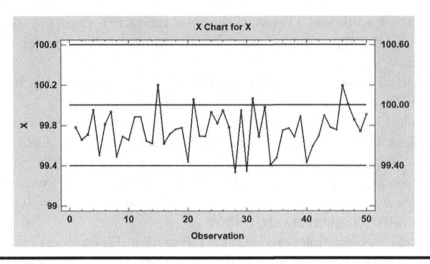

Figure 3.1   Traditional X chart, one-sigma process shift

When *A* is 2 times sigma as recommended, the absolute value of the slopes of the V-mask arms is:

$$K = \frac{\delta\sigma}{2} \text{ or alternatively } \frac{\Delta}{2}$$

Recall from the previous example that the nominal is 100, the specification limits are 99 and 101, respectively, and the process standard deviation is 0.2. We want to detect a shift of 1 process standard deviation with a false alarm risk of 0.01 and a negligible beta risk of not detecting the shift. Figure 3.1 from StatGraphics shows that the traditional X chart detects this on about the 15th observation.

To set up the CUSUM chart,

$$d = -\left(\frac{2}{1^2}\right)\ln\left(\frac{0.01}{2}\right) = 10.60$$

$$K = \frac{1 \times 0.2}{2} = 0.1$$

Figure 3.2 from StatGraphics shows that the CUSUM chart detects the out of control situation on the eighth rather than the 15th observation as the first goes outside the V-mask when the eighth measurement is applied. It also shows that the absolute value of the slope of the V-mask ought to be simply *H/d*, and 1.058 divided by 10.6 is in fact roughly 0.1.*

---

* The (8, −1.10) point is slightly uncertain as it relies on aligning the cross-hairs of the location feature on the V-mask precisely, which is not really possible.

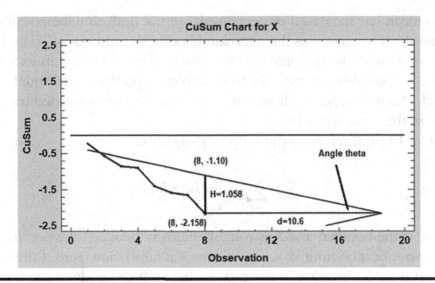

Figure 3.2   CUSUM chart for individuals

Note the out of control signal for the first point, which shows that this chart detected the condition of interest after only eight measurements. *Out of control signals appear for trailing points as the V-mask moves forward with new measurements, as opposed to at the most recent point where we expect them for traditional SPC.*

We have seen so far that this chart works very well because it detects the 1-sigma process shift more rapidly than the traditional control chart, and this is because it accumulates (as in "cumulative sum") the deviations from nominal. It will work for individual measurements or sample averages, although the samples must be of identical size for the method shown above to work. This makes it ideal, at least in theory, for processes with no prior history but for which the desired nominal and standard deviation can be specified.

Deployment of this kind of chart to a spreadsheet is, on the other hand, extremely difficult. Remember that we want something into which the production worker needs to enter only the measurements to get a simple visual signal of an out of control situation as opposed to having to interpret a V-mask. The latter can probably be developed in Excel, but it needs to advance automatically every time a new measurement or sample is entered. A tabular variant of CUSUM (Montgomery, 1991, 290–295 and also StatGraphics' procedure documentation) is, on the other hand, suited perfectly for spreadsheets.

Tabular CUSUM

Montgomery (1991, 290) points out that tabular CUSUM was designed as a one-sided rather than two-sided test, i.e. a test to determine whether the process mean has increased or decreased, but not both simultaneously. That is, a one-sided test assesses the alternate hypothesis (and out of control condition) "more than the nominal" or "less than the nominal" on an exclusive or basis. A two-sided test assesses the alternate hypothesis "not equal to the nominal." Nothing stops us, however, from running two one-sided tests side by side in the same spreadsheet.

Tabular CUSUM relies on the following equation:

$$S_i = \sum_{j=1}^{i} \left[\bar{x}_j - \left(\mu_0 + K \right) \right]$$

where μ_0 is the nominal and K is roughly halfway between the nominal and the specification limit. S_i resets to zero if it falls below zero if the one-sided test is for an increase in the process mean. It can address the two-sided application, and in a recursive form as depicted in Equation Set 3.2. "Recursive" means that, for example, a spreadsheet cell can obtain the necessary value from the one directly above it as opposed to having to perform an entire summation.

$$S_{H,i} = \max \left[0, \bar{x}_i - \left(\mu_0 + K \right) + S_{H,i-1} \right] \quad \text{where } S_{H,0} = 0 \qquad (3.2a)$$

$$S_{L,i} = \max \left[0, \left(\mu_0 - K \right) - \bar{x}_i + S_{L,i-1} \right] \quad \text{where } S_{L,0} = 0 \qquad (3.2b)$$

S_H will increase steadily if the measurement or sample average runs consistently above the nominal plus K, and S_L will increase steadily if the measurement or sample average runs consistently below the nominal minus K. Note that both of these quantities are positive, and reset to 0 if they become negative.

The decision interval H, which is the distance from the center line of a CUSUM chart to each line of its V-mask, is

$$H = 2d\sigma \times \tan(\theta) \quad \text{where } d = \left(\frac{2}{\delta^2} \right) \ln \left(\frac{1-\beta}{\frac{\alpha}{2}} \right)$$

where $\tan(\theta)$ comes from Table 7.2, Cumulative-Sum Control Chart Parameters, from Montgomery (1991, 289). The inputs to this table are (1) the process shift δ (standard deviations) to be detected and (2) the desired average run length (ARL) when the process is in control.

How did we get to ARL from specified alpha and beta risks? Montgomery (1991, 287) states that, in practice, the actual risks do not match the specified ones, so practitioners use the desired ARL when the process is in control to reflect the false alarm risk (alpha). The idea is then to minimize the ARL, and therefore the beta risk, for the specified process shift. This chapter will show that Excel's Solver feature can do this and returns results almost identical to those of StatGraphics.

In addition, $H=h\sigma$, so $b=2d\times\tan(\theta)$, which is admittedly not very helpful if we don't have the table. The StatGraphics documentation adds, however, that the average run length is:

$$\frac{1}{ARL} = \frac{1}{ARL_+} + \frac{1}{ARL_-}$$

$$ARL_{one\_sided} = \frac{\exp(-2\Delta b) + 2\Delta b - 1}{2\Delta^2}$$

where $\Delta = \delta - k$ for ARL_+ and $-\delta - k$ for ARL_- and $b=h+1.166$. (As δ and k are both in number of standard deviations, K will then be $k\sigma_{process}$.) This is Siegmund's formula (Maghsoodloo, 2013), and it is indeterminate when $\Delta=0$, i.e. $k=\delta$ for ARL_+ and $k=-\delta$ for ARL_-. This can, however, be dealt with through application of L'Hôpital's Rule. Take the first derivatives of the numerator and denominator:

$$\lim_{\Delta\to 0} \frac{\exp(-2\Delta b) + 2\Delta b - 1}{2\Delta^2} = \frac{-2b\times\exp(-2\Delta b) + 2b}{4\Delta}$$

This is still indeterminate at $\Delta=0$, so take the second derivatives to get:

$$\lim_{\Delta\to 0} \frac{\exp(-2\Delta b) + 2\Delta b - 1}{2\Delta^2} = \frac{4b^2\times\exp(-2\Delta b)}{4}$$

This gives us Equation Set 3.3 and therefore two equations for two unknowns.

$$ARL(0) = \frac{4b^2\times\exp(-2\Delta b)}{4} = b^2 \tag{3.3a}$$

$$\text{ARL}(\delta) = \frac{\exp(-2\Delta b) + 2\Delta b - 1}{2\Delta^2} \qquad (3.3b)$$

Decision Interval and K

Equation set 3.3 allows us to obtain the decision interval parameter H, along with K, even if we don't have the necessary table. Recall that the inputs are (1) the process shift δ (standard deviations) to be detected and (2) the average run length (ARL) when the process is in control. The deliverables are H and K, and we obtain them as follows from Equation Set 3.4 noting that, for a two-sided test, $\text{ARL}_- = \text{ARL}_+$. Minimize

$$\text{ARL}_+(\delta) = \frac{\exp(-2\Delta b) + 2\Delta b - 1}{2\Delta^2} = \frac{\exp(-2(\delta - k)b) + 2(\delta - k)b - 1}{2(\delta - k)^2} \qquad (3.4a)$$

subject to the condition that ARL(0) equals the specified ARL.

$$\text{ARL}(0) = \frac{\exp(-2(-k)b) + 2(-k)b - 1}{2(-k)^2} = \frac{\exp(2kb) - 2kb - 1}{2k^2} \qquad (3.4b)$$

Recall that $\Delta = \delta - k$ for ARL_+ and $-\delta - k$ for ARL_- and $b = h + 1.166$. We can use Excel's Solver feature to minimize Equation 3.4a subject to the constraint given by 3.4b to deliver k and b, and then $h = b - 1.166$.

Montgomery (1991, 291–292) gives an example in which the objective is to detect a process shift of one standard deviation with an ARL of roughly 500 when no process shift has occurred. The use of Table 7.2 from Bowker and Lieberman delivers $H = 4.986$ and $K = 0.5$. StatGraphics returns $H = 5.07$ and $K = 0.5$, with an ARL of 10.52 for the 1-sigma process shift. The next question is as to whether we can perform the same calculation in Excel.

Figure 3.3 shows the problem statement in Excel. Note that, as the total ARL is to be 500, we must use 1000 as the argument for ARL because the subsequent equations calculate the one-sided ARL. The corresponding argument for StatGraphics' Design CUSUM Chart is, however, the two-sided risk. That is, we use 500 in StatGraphics and 1000 in the Excel procedure shown below.

$$\frac{1}{\text{ARL}} = \frac{1}{500} = \frac{1}{\text{ARL}_+} + \frac{1}{\text{ARL}_-} = \frac{1}{1000} + \frac{1}{1000}$$

- ■ Cell B6: =B2-B5
- ■ Cell B7: =(EXP(2*B$5*B$4)-2*B$5*B$4-1)/(2*B$5^2)

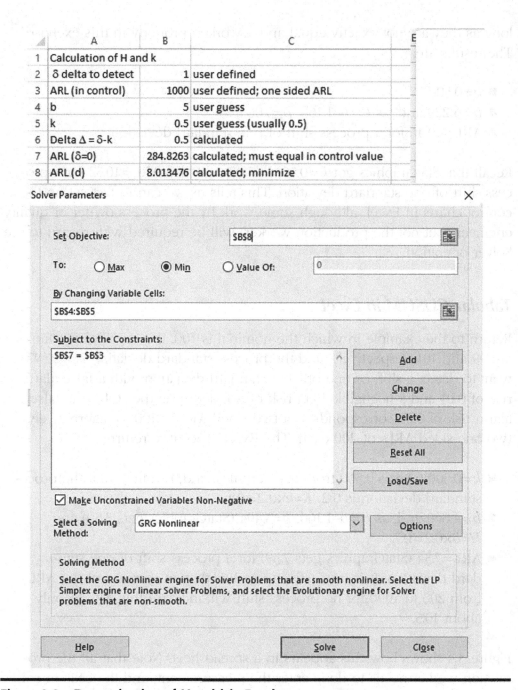

	A	B	C	E
1	Calculation of H and k			
2	δ delta to detect	1	user defined	
3	ARL (in control)	1000	user defined; one sided ARL	
4	b	5	user guess	
5	k	0.5	user guess (usually 0.5)	
6	Delta Δ = δ-k	0.5	calculated	
7	ARL (δ=0)	284.8263	calculated; must equal in control value	
8	ARL (d)	8.013476	calculated; minimize	

Solver Parameters ✕

Se**t** Objective: $B$8

To: ○ **M**ax   ⦿ Mi**n**   ○ **V**alue Of: 0

By Changing Variable Cells:

$B$4:$B$5

Su**b**ject to the Constraints:

$B$7 = $B$3

    Add

    Change

    Delete

    Reset All

    Load/Save

☑ Ma**k**e Unconstrained Variables Non-Negative

S**e**lect a Solving Method: GRG Nonlinear   ⌄   **O**ptions

Solving Method

Select the GRG Nonlinear engine for Solver Problems that are smooth nonlinear. Select the LP Simplex engine for linear Solver Problems, and select the Evolutionary engine for Solver problems that are non-smooth.

Help   **S**olve   Cl**o**se

Figure 3.3 Determination of *H* and *k* in Excel

■ Cell B8: =(EXP(-2*B$6*B$4)+2*B$6*B$4-1)/(2*B$6^2)

Use Excel's Solver feature to minimize cell B8, the ARL at the specified process shift δ, subject to the constraint that cell B7 equal the specified ARL for $\delta = 0$. As the function in cell B8 is undefined when $\delta = k$, we could add another constraint that the two not be equal, but the function will work as long as they are not exactly equal and it worked properly in this exercise. The results are:

- $k=0.500$
- $b=6.229$ and, as $h=b-1.166$, $h=5.063$.
- ARL = 10.46 for a process shift of one standard deviation

Recall that StatGraphics got $k=0.50$, $h=5.07$, and also ARL = 10.52 for a process shift of one standard deviation. This tells us we can in fact set up these control charts in Excel, although some work by the process owner or quality engineer (but not the production worker) will be required with regard to the Solver operation.

Tabular CUSUM in Excel

Return to the example in which the nominal is 100, the specification limits are 99 and 101, respectively, and the process standard deviation is 0.2. We want to detect a shift of one process standard deviation with a false alarm risk of 0.01 and a negligible beta risk of not detecting the shift. The false alarm risk of 0.01 corresponds to a two-sided ARL of 100, or alternatively two one-sided ARLs of 200 each. The Excel procedure returns:

- $k=0.500$ (versus 0.500 from StatGraphics) and, recalling that the process standard deviation is 0.2, $K=k \times 0.2 = 0.1$.
- $b=4.660$ and, as $h=b-1.166$, $h=3.494$ (StatGraphics gets 3.50) and $H=h \times 0.2 = 0.7$.
- ARL = 7.34 (StatGraphics gets 7.39) for a process shift of one standard deviation. It is instructive to realize that an increase of the ARL from 200 to 1000 for no process shift will increase this ARL to only about 10.5.

Figure 3.4 shows how this appears in a spreadsheet. Note that all the production worker needs to do is enter the measurement, and the cells in columns G and H do the rest.

- Cell G12: =MAX(0,F12-(F\$5+F\$7)+G11), i.e. X-(nominal+K)+$S_{H,i-1}$
- Cell H12: =MAX(0,(F\$5-F\$7)-F12+H11), i.e. (nominal-K)-$S_{L,i-1}$
- Conditional formatting turns a cell red if it exceeds the value H in F6.

⊿	E	F	G	H
1	Tabular CUSUM Example			
2	h	3.494	user entry	
3	k	0.5	user entry	
4	sigma	0.2	user entry	
5	nominal	100	user entry	
6	H	0.6988	calculated, = h*sigma	
7	K	0.1	calculated, = k*sigma	
8				
9				
10	i	X	S_H	S_L
11	0		0	0
12	1	99.783	0.000	0.117
13	2	99.661	0.000	0.356
14	3	99.707	0.000	0.550
15	4	99.953	0.000	0.496
16	5	99.500	0.000	0.896
17	6	99.812	0.000	0.984
18	7	99.936	0.000	0.948
19	8	99.490	0.000	1.357
20				

Figure 3.4 Tabular CUSUM in a spreadsheet

StatGraphics gets similar results as shown in Figure 3.5. The documentation explains that C_+ and C_- are the values of the positive and negative CUSUMs, respectively, and note that they match those in the Excel spreadsheet. The scale CUSUM results are used to monitor process variability.

StatGraphics also (Figure 3.6) generates what Montgomery (1991, 294) calls a CUSUM status chart.

- The individual point is the deviation of the measurement from the nominal. The first point is −0.217, or 99.783–100.

CuSum Individuals Chart Report
All Observations
X = Excluded   * = Beyond Limits

C = Mean Cusum S = Scale Cusum

Observation	C+	N+	C-	N-	S+	N+	S-	N-
1	0.0	0	0.117	1	0.12932	1	0.0	0
2	0.0	0	0.356	2	1.0045	2	0.0	0
3	0.0	0	0.549	3	1.6173	3	0.0	0
4	0.0	0	0.496	4	0.15099	4	0.46628	1
5	0.0	0	*0.896	5	1.8262	5	0.0	0
6	0.0	0	*0.984	6	1.7489	6	0.0	0
7	0.0	0	*0.948	7	0.51448	7	0.23443	1
8	0.0	0	*1.358	8	2.2347	8	0.0	0

Figure 3.5   Tabular CUSUM in StatGraphics

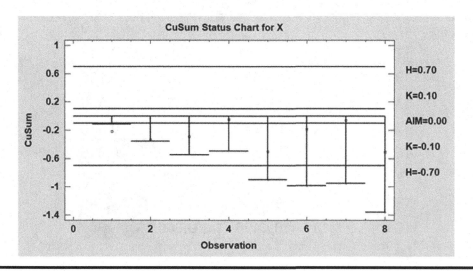

Figure 3.6   CUSUM status chart

■ The vertical bars are the cumulative sums, and the positive ones will be above the center line.
■ The horizontal lines at ±0.10 from the center lines are the slack or reference values.
■ The lines at ±0.70 represent the decision interval.

Excel can develop bar charts, and also bar charts with lines to mark the center line, slack or reference values, and the decision interval if this will help the production worker identify a situation that requires corrective action as determined here with the fifth observation. Figure 3.7 shows a

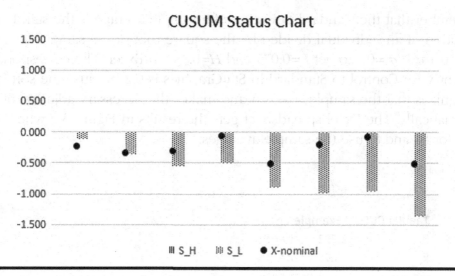

Figure 3.7   CUSUM status chart in Excel

combination bar (for CUSUMs) and point (for deviations from nominal) chart to demonstrate that this can be done, although I could not make the points line up with the centers of the bars. Remember, however, that the conditional formatting of the cells that contain the CUSUMs should alert the production worker of a process shift, so the chart might not even be necessary.

Tabular CUSUM for Samples

Tabular CUSUM also works for samples, although it looks like the samples must all be the same size in contrast to charts for the standard normal deviate that accommodate varying sample sizes. Continue with the example for which the nominal is 100 and the required process performance index is 5/3 which, given a specification width of 2, requires a standard deviation of 0.2. What happens if the sample size is 4, and we want to detect a process shift of 1.5 standard deviations? Also assume we want a false alarm risk equal to that of a Shewhart chart (0.0027), which means the two-sided average run length is 370. Remember that StatGraphics will use this and convert it into the necessary one-sided ARL of 740, but we must use 740 for Excel's Solver feature. The results are as follows:

- $k=0.75$ (Excel solver and StatGraphics)
- $h=3.323$ (Excel) and 3.34 (StatGraphics)
- ARL at $\delta=1.5$ is 5.10 from Excel, 5.18 from StatGraphics

Remember that the standard deviation of a sample average is the standard deviation of an individual divided by the square root of the sample size, so we must use $\sigma = 0.1$ to get $K = 0.075$ and $H = 0.332$ in Excel. The corresponding entry for Control to Standard in StatGraphics is 0.2 because the software recognizes that the sample size is 4 and makes the necessary adjustment automatically. The Excel spreadsheet gets the results in Figure 3.8 where the cells for S_H and S_L use the sample averages.

	E	F	G	H	I	J	K
	Tabular CUSUM Example						
	h	3.323	user entry				
	k	0.75	user entry				
	sigma	0.1	user entry				
	nominal	100	user entry				
	H	0.3323	calculated, = h*sigma				
	K	0.075	calculated, = k*sigma				
	X1	X2	X3	X4	x_bar	S_H	S_L
						0	0
1	100.203	100.303	99.936	99.934	100.094	0.0191	0.0000
2	100.055	99.826	99.740	99.893	99.879	0.0000	0.0465
3	100.198	100.070	100.262	100.162	100.173	0.0980	0.0000
4	100.020	99.809	100.219	100.248	100.074	0.0968	0.0000
5	99.949	100.169	99.827	100.072	100.004	0.0262	0.0000
6	100.096	99.750	100.037	99.904	99.947	0.0000	0.0000
7	99.767	100.093	99.934	100.205	100.000	0.0000	0.0000
8	99.905	99.602	100.060	99.625	99.798	0.0000	0.1270
9	100.121	100.143	100.426	100.113	100.201	0.1258	0.0000
10	100.109	100.050	100.025	100.062	100.061	0.1122	0.0000
11	100.103	100.251	100.078	100.315	100.187	0.2239	0.0000
12	99.666	99.914	100.035	99.938	99.888	0.0372	0.0367
13	100.013	100.002	99.928	100.150	100.023	0.0000	0.0000
14	99.646	100.219	100.054	99.442	99.840	0.0000	0.0847
15	100.050	100.371	99.928	99.874	100.056	0.0000	0.0000
16	99.698	99.935	99.687	99.969	99.822	0.0000	0.1027
17	100.102	99.754	100.113	100.090	100.015	0.0000	0.0131
18	99.597	99.798	99.774	100.192	99.840	0.0000	0.0980
19	99.764	100.168	99.974	100.455	100.090	0.0150	0.0000
20	100.134	99.888	100.169	100.055	100.061	0.0013	0.0000

Figure 3.8   Tabular CUSUM in Excel for sample averages

StatGraphics gets similar results, as shown in Figure 3.9, and the status chart appears in Figure 3.10.

This chapter has shown so far, then:

■ CUSUM is well suited for control-to-standard applications in which the user specifies the nominal and (based on the specification width and required process performance index) standard deviation.
■ Tabular CUSUM is easily deployable in a spreadsheet, and Excel's Solver feature can obtain the required parameters (h and k) from the required ARL for a process shift of zero and minimization of the ARL for the process shift that is to be detected.
■ Tabular CUSUM will work for individual measurements, which is especially useful when we seek to detect relatively small discrepancies between the process mean and the nominal, and also for samples. The limitation on the latter application appears to be that the samples must be of identical sizes.

C = Mean Cusum

Subgroup	C+	N+	C-	N-	Range
1	0.019062	1	0.0	0	0.36881
2	0.0	0	0.04645	1	0.31577
3	0.098006	1	0.0	0	0.1916
4	0.096845	2	0.0	0	0.43964
5	0.026191	3	0.0	0	0.34146
6	0.0	0	0.0	0	0.34642
7	0.0	0	0.0	0	0.43872
8	0.0	0	0.12704	1	0.45827
9	0.12584	1	0.0	0	0.31257
10	0.11219	2	0.0	0	0.083282
11	0.22393	3	0.0	0	0.23667
12	0.037213	4	0.03672	1	0.36819
13	0.0	0	0.0	0	0.22227
14	0.0	0	0.08466	1	0.77635
15	0.0	0	0.0	0	0.49755
16	0.0	0	0.10267	1	0.28202
17	0.0	0	0.013118	2	0.3591
18	0.0	0	0.098024	3	0.59527
19	0.014955	1	0.0	0	0.69069
20	0.0013246	2	0.0	0	0.2807

Figure 3.9   Tabular CUSUM (StatGraphics) for samples of four

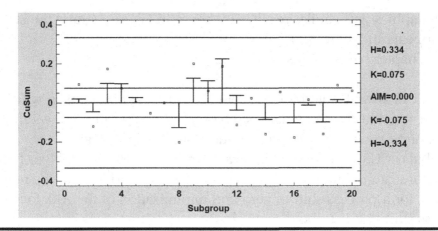

Figure 3.10   CUSUM status chart (StatGraphics), samples of four

The exponentially weighted moving average (EWMA) chart also is useful for detecting relatively small discrepancies between the process mean and the nominal, and is easier to use. This chapter will address EWMA in detail later.

Cuscore

Box and Luceño (1997, Chapter 10) add that CUSUM charts are very capable of detecting signals that are hidden in noise. A *Cuscore* (cumulative score) statistic is, however, customized to detect a *specific* kind of undesirable process change that would otherwise remain hidden in white noise. Page 234 of the reference describes a customized Cuscore statistic that adds a sine function for the express purpose of detecting harmonic cycling.

   Another application multiplies a traditional CUSUM statistic by time to detect an unusually rapid increase in tool wear. Feigenbaum (1991, 423–424) discusses the issue of control charts with slanting control limits for processes with tool wear of a known rate. Suppose, for example, a process has specification limits of [98,102] and a standard deviation of 0.25, which makes it an 8-sigma process. Tool wear results, however, in a 0.02 reduction of dimension size with each successive part. The control limits for the X chart for individuals are therefore for the ith part (given there is no tool wear for the first part):

$$\left(100 - 0.02(i - 1)\right) \pm 3 \times 0.25$$

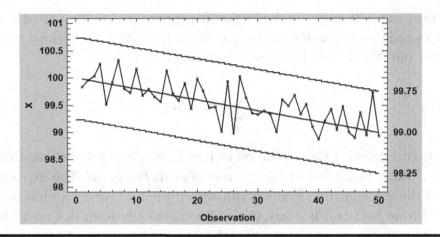

Figure 3.11   Individuals chart for a process with a known tool wear rate

StatGraphics generates Figure 3.11 given a starting dimension of 100, a wear rate of −0.02, and a standard deviation of 0.25. StatGraphics can also estimate the tool wear rate from data if it is not given.

This is traditional SPC for a process that involves tool wear. Cuscore can, however, detect problems more rapidly, as discussed in the next section.

CUSUM versus Cuscore

Recall that the quantity plotted for the *i*th measurement or sample average is, per Montgomery (1991, 280–285):

$$S_i = \sum_{j=1}^{i} \left(\overline{x}_j - \mu_{\text{target}} \right) = \sum_{j=1}^{i} \text{dnom}_j$$

Another way to say this is that, if E is the expected value of the sample average or individual,

$$S_i = \sum_{j=1}^{i} \left(\overline{x}_j - E\left(\overline{x}_j \right) \right)$$

In the case of tool wear with rate beta, the expected value of the *t*th (*t*=time or amount of use) average or individual is the initial starting dimension minus beta times the number of units produced. We will use *t*=0 to *i* because there is no wear for the first part, and let *y*-bar at time *t* be the tool wear as estimated from the sample statistics. If, for example, the starting dimension was 100 mils, the dimension is proportional to the tool size, and the estimated dimension for the *t*th part is 99 mils, then the estimated tool wear is 1 mil. The CUSUM statistic is computed as follows:

$$S_i = \sum_{t=0}^{i} \left(\bar{y}_t - \beta t \right)$$

The Cuscore statistic Q is, on the other hand, the sum of deviations from the expected value *multiplied by the number of parts produced*. The expected value of these summation terms is still zero if the tool wear rate has not changed from beta but, if it has, the additional weight from this multiplication will result in a more rapid signal than would be obtained from ordinary CUSUM.

$$Q_i = \sum_{t=0}^{i} \left(\bar{y}_t - \beta t \right) t$$

Box and Ramirez (1992) provide additional details on this method and add that it gives primarily a subjective, albeit very obvious, signal of an undesirable change in the process. They add, however, that significance tests can be performed if desired.

An attempt was made to reproduce results similar to those in the Box and Luceño (1997, p. 237) reference for an increase in tool wear rate. A process has a starting dimension of 100 mils, random variation of 0.02 mils, and an expected tool wear rate of 0.02 mils per part. Fifty parts were simulated whose expected dimension is 100 mils minus 0.02 times the part number, with the first part being zero as there is no tool wear. The estimated tool wear is therefore 100 mils minus the current part dimension; the data appear in Chapter_3_Cuscore in the Excel file that accompanies the book. The first simulation uses a wear rate of 0.02 mils for all 50 parts for which the Cuscore for individuals is:

$$Q_i = \sum_{t=0}^{i} \left(y_t - 0.02t \right) t$$

Figure 3.12 shows the result when the tool wear rate is as expected. The fact that successive terms are multiplied by the number of parts suggests that the magnitude of the swings will increase, but the swings oscillate around zero for this simulation.

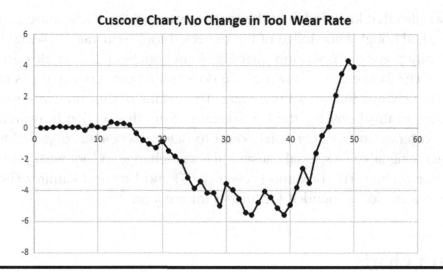

Figure 3.12   Cuscore chart for tool wear, no change in rate

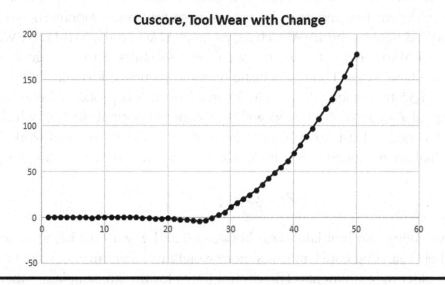

Figure 3.13   Cuscore chart for tool wear, rate increases on 26th part

Now suppose that the tool wear rate increases from 0.02 to 0.025 mils per part, and this change begins on the 26th part. Figure 3.13 looks similar to the one in the reference, which begins with a wear rate of 1 and increases to 1.25 on the 11th part. The problem becomes obvious fairly quickly, and close examination of the chart shows that the change happened between the 25th and 30th part. Note also the difference between the scales of the ordinates when the tool wear is as expected (Figure 3.12) and when it has increased (Figure 3.13).

Note also that knowledge of the random variation or white noise is not required, although knowledge of the expected tool wear rate is. While it is apparently possible to perform quantifiable hypothesis tests per Box and Ramirez (1992), the chart shown above does not require control limits to reveal the change in the tool wear rate. The details of Cuscore are beyond the scope of this book but the key takeaway from this section is awareness of its existence and its enormous power to detect *specific* (as opposed to generic) assignable or special causes such as a change in tool wear rate or harmonic cycling. The Box and Luceño (1997) and Box and Ramirez (1992) references are recommended for further information.

EWMA Charts

The exponentially weighted moving average is similar to CUSUM but it discounts older measurements in favor of more recent ones. Montgomery (1991, 299–306) describes the EWMA chart, as does AIAG (2005, 111–112). EWMA is, like CUSUM, better at detecting small process shifts but not as good at detecting large ones. The test statistic uses the recursive formula shown in Equation 3.5 that is ideal for use in a spreadsheet. It is probably better to use the capital Z to avoid confusion with the standard normal deviate, which the EWMA is not, at least not in the absence of conversion. This will work for individual measurements or sample averages, as shown in Equation 3.5.

$$Z_i = \lambda \bar{x}_i + \left(1 - \lambda\right) Z_{i-1} \tag{3.5}$$

The weighting constant lambda is between 0 and 1, with 0.2 being typical. A higher weighting constant gives more weight to recent measurements and less to older measurements. The control limits for the *i*th sample are then, for a target or nominal of μ_0 and a sample size of n, given by Equation Set 3.6.

$$\mu_0 \pm 3\sigma \sqrt{\frac{\lambda}{\left(2 - \lambda\right)n}\left(1 - \left(1 - \lambda\right)^{2i}\right)} \tag{3.6a}$$

The limit as i approaches infinity (or, for practical purposes, 20 or so) is:

$$\mu_0 \pm 3\sigma \sqrt{\frac{\lambda}{\left(2 - \lambda\right)n}} \tag{3.6b}$$

EWMA for Individuals

Return to the example in which the nominal is 100, the specification limits are 99 and 101 respectively, and the process standard deviation is 0.2. Figure 3.14 shows how easy this is to address in Excel.

■ Cell G5=$B$3*F5+(1-$B$3)*G4 where B3 contains the weighting constant of 0.2. The result is therefore 0.2×99.783+(1-0.2)×100=99.957

	E	F	G	H	I
1	Simulated process shift = 1 standard deviation				
2					
3	i	X	Z	LCL	UCL
4	0		100		
5	1	99.783	99.957	99.880	100.120
6	2	99.661	99.897	99.846	100.154
7	3	99.707	99.859	99.828	100.172
8	4	99.953	99.878	99.818	100.182
9	5	99.500	99.803	99.811	100.189
10	6	99.812	99.804	99.807	100.193
11	7	99.936	99.831	99.804	100.196
12	8	99.490	99.763	99.803	100.197
13	9	99.691	99.748	99.802	100.198
14	10	99.654	99.729	99.801	100.199
15	11	99.884	99.760	99.801	100.199
16	12	99.882	99.785	99.800	100.200
17	13	99.645	99.757	99.800	100.200
18	14	99.618	99.729	99.800	100.200
19	15	100.201	99.823	99.800	100.200
20	16	99.621	99.783	99.800	100.200
21	17	99.713	99.769	99.800	100.200
22	18	99.766	99.768	99.800	100.200
23	19	99.780	99.771	99.800	100.200
24	20	99.442	99.705	99.800	100.200
25	21	100.057	99.775	99.800	100.200
26	22	99.695	99.759	99.800	100.200

Figure 3.14   EWMA for individual measurements

- Cell H5 = $B$1-3*$B$2*SQRT((($B$3/(2-$B$3))*(1-(1-$B$3)^(2*$E5)))) where B1 contains the nominal, B2 the standard deviation of 0.2, and B3 the weighting factor. Conditional formatting turns this cell red if it is greater than Z, i.e. Z is below the lower control limit.
- Cell I5 = $B$1+3*$B$2*SQRT((($B$3/(2-$B$3))*(1-(1-$B$3)^(2*$E5)))). Conditional formatting turns this cell red if it is less than Z, i.e. Z exceeds the upper control limit. Both control limits converge to the limiting case depicted by Equation (3.6b) and also as determined by StatGraphics.

Note that the out of control condition is detected with the fifth measurement under this approach as shown in Figure 3.15.*

Figures 3.16 and 3.17 show the respective control charts from Excel and StatGraphics.

EWMA charts therefore perform the same task as CUSUM charts, i.e. the detection of relatively small discrepancies between the process mean and the nominal. They are computationally much simpler to deploy in a spreadsheet and they also meet the requirement that the production worker needs to do nothing more than enter measurements to get actionable information about the status of the process. The process owner or quality engineer can easily include the charts, but conditional formatting of the cells eliminates the need to examine the charts to identify an out of control condition.

Selection of Method: Traditional SPC, CUSUM, or EWMA?

The most likely objective of short-run SPC is to test the hypothesis that a startup process, i.e. one with no prior history, or a startup after a changeover, is centered on its nominal. PRE-Control tests this assumption by requiring the production of five successive parts in the green zone that surrounds the nominal.

Montgomery (1991, 306) states that CUSUM and EWMA are better at detecting small process shifts than traditional SPC, but not as good at detecting larger process shifts. This reference also recommends that the weighting factor lambda be between 0.05 and 0.25 for EWMA. Smaller values of lambda, which give more weight to the previous measurements or samples, are better at detecting smaller shifts. This suggests that if the principal risk consists of a small discrepancy between the process mean and the nominal, CUSUM or EWMA would be the best choice. If the principal risk involves a large discrepancy, then traditional SPC is best.

---

* Select initialization with variable limits to get StatGraphics to use Equation (3.6a).

EWMA Individuals Chart Report		
All Observations		
X = Excluded * = Beyond Limits		
Observation	EWMA	MR(2)
1	99.957	
2	99.897	0.122
3	99.859	0.046
4	99.878	0.246
5	* 99.802	0.453
6	* 99.804	0.312
7	99.831	0.124
8	* 99.763	0.446
9	* 99.748	0.201
10	* 99.729	0.037
11	* 99.76	0.23
12	* 99.785	0.002
13	* 99.757	0.237
14	* 99.729	0.027
15	99.823	0.583
16	* 99.783	0.58
17	* 99.769	0.092
18	* 99.768	0.053
19	* 99.771	0.014
20	* 99.705	0.338
21	* 99.775	0.615
22	* 99.759	0.362
23	* 99.746	0.003
24	* 99.783	0.241

Figure 3.15   EWMA, StatGraphics

Average Run Length, Traditional SPC

The chance that a point will be inside the traditional 3-sigma limits of a Shewhart chart, which is the consumer's (beta) risk of not detecting a specified process shift, is, where Φ is the cumulative standard normal distribution function,

$$\beta(\mu) = \Phi\left(\frac{\text{UCL} - \mu}{\frac{\sigma}{\sqrt{n}}}\right) - \Phi\left(\frac{\text{LCL} - \mu}{\frac{\sigma}{\sqrt{n}}}\right)$$

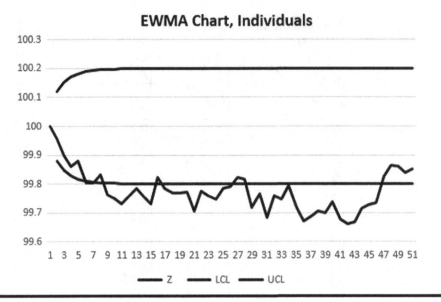

Figure 3.16   EWMA control chart, Excel

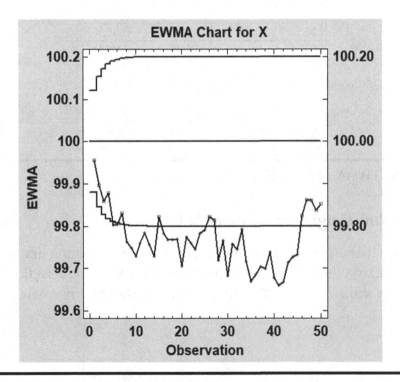

Figure 3.17   EWMA control chart, StatGraphics

Noting that the control limits are the nominal ±3 times the sample standard deviation,

$$\beta(\mu) = \Phi\left(\dfrac{\mu_0 - \mu + 3\dfrac{\sigma}{\sqrt{n}}}{\dfrac{\sigma}{\sqrt{n}}}\right) - \Phi\left(\dfrac{\mu_0 - 3\dfrac{\sigma}{\sqrt{n}} - \mu}{\dfrac{\sigma}{\sqrt{n}}}\right)$$

$$\beta(\mu) = \Phi\left(\dfrac{\mu_0 - \mu}{\dfrac{\sigma}{\sqrt{n}}} + 3\right) - \Phi\left(\dfrac{\mu_0 - \mu}{\dfrac{\sigma}{\sqrt{n}}} - 3\right)$$

This can be simplified, and generalized even further, as follows, noting that the standard normal deviate is, when δ is given in sample standard deviations,

$$z = \delta = \dfrac{\mu - \mu_0}{\dfrac{\sigma}{\sqrt{n}}}$$

$$\beta(\delta) = \Phi(-\delta + 3) - \Phi(-\delta - 3) = \left(1 - \Phi(\delta - 3)\right) - \left(1 - \Phi(\delta + 3)\right)$$

Equation set 3.7 provides the consumer's risk and ARL for a process shift of δ standard deviations.

$$\beta(\delta) = \Phi(\delta + 3) - \Phi(\delta - 3) \tag{3.7a}$$

$$\mathrm{ARL}(\delta) = \dfrac{1}{1 - \beta(\delta)} = \dfrac{1}{1 - \Phi(\delta + 3) + \Phi(\delta - 3)} \tag{3.7b}$$

If, for example, there is no process shift, the chance that the point will be inside the control limits is $\Phi(3) - \Phi(-3) = 0.9973$, so the false alarm risk of a point being outside the control limits is 0.0027 and the ARL is 370.4.

What, on the other hand, is the chance of detecting a 1.5-sigma process shift (which results in 3.4 defects per million opportunities in a Six Sigma process) if the sample size is 4? The shift is three sample standard deviations, i.e. the square root of four times sigma, and $\Phi(6) - \Phi(0) = 0.5$. The average run length under these conditions is 2.

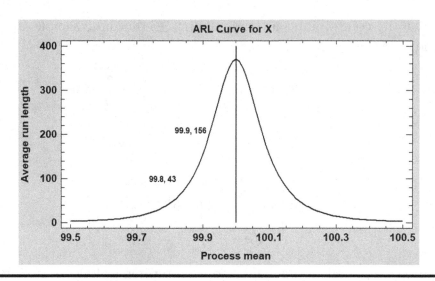

Figure 3.18   ARL, Shewhart chart for individuals, sigma=0.2

Figure 3.18 shows a StatGraphics graph of the ARL for a Shewhart control chart for a chart for individuals with a nominal of 100 and standard deviation of 0.2. The process shifts are therefore 0.5-sigma and 1-sigma, respectively, at 99.9 and 99.8. The indicated ordinates are slightly uncertain as they are obtained by lining up a crosshair with the Locate feature, i.e. they depend on how closely the user can align them to the graph.

$$\mathrm{ARL}(0.5) = \frac{1}{1 - \beta(0.5)} = \frac{1}{1 - \Phi(0.5 + 3) + \Phi(0.5 - 3)} = 155.2$$

$$\mathrm{ARL}(1) = \frac{1}{1 - \beta(1)} = \frac{1}{1 - \Phi(1 + 3) + \Phi(1 - 3)} = 43.9$$

This shows that it is extremely easy to calculate the ARL for any process shift, as measured in sample standard deviations, for a traditional Shewhart control chart.

Average Run Length, CUSUM

Recall Equation 3.4b, which delivers the ARL for a tabular CUSUM control scheme:

$$\mathrm{ARL}(\delta) = \frac{\exp(-2\Delta b) + 2\Delta b - 1}{2\Delta^2} \qquad (3.4b)$$

where $\Delta=\delta-k$ for ARL_+ and $-\delta-k$ for ARL_- and $b=h+1.166$. This chapter has already shown that k and b can be optimized for any δ in Excel subject to the condition that $ARL(0)$ be a specified number, and StatGraphics also will perform this task.

Suppose, for example, that we want $ARL(0)$ to be 370 to reflect the false alarm risk of a traditional Shewhart chart. The previous section showed that $ARL(0.5)=155$ and $ARL(1.0)=44$ for the traditional Shewhart chart for individuals. What are these ARLs if a CUSUM scheme is optimized to minimize the ARL for these process shifts?

■ For $\delta=0.5$, StatGraphics finds $k=0.25$, $b=8.01$, and $ARL=28.8$ which is substantially less than 155. The Excel solver method gets almost identical results, but remember that $ARL(0)$ must be specified as the one-sided ARL, which is 740. This means that tabular CUSUM will detect a 0.5-sigma process shift much more rapidly than a traditional X chart.
■ For $\delta=1$, StatGraphics finds $k=0.5$, $b=4.77$, and $ARL=9.33$ which is substantially less than 44. The Excel method yields almost identical results.

This shows that we can compare, for any given process shift, the ARL from a traditional Shewhart chart to that of a tabular CUSUM control method.

Average Run Length, EWMA

The ARL for an EWMA chart is far less accessible in Excel, and may not even be possible to obtain without extensive programming in Visual Basic for Applications (VBA). Montgomery (1991, 305) cites Crowder (1987 and 1989), and StatGraphics' documentation for EWMA also cites Crowder (1987). The Crowder references provide the following equation, for a process shift of u as measured in process standard deviations, and the StatGraphics' documentation indicating that the control limits are ±3, respectively, i.e. traditional 3-sigma Shewhart control limits. $L(u)$ is the ARL for the indicated shift.

$$L(u) = 1 + \frac{1}{\lambda}\int_{LCL}^{UCL} L(y)\phi\left(\frac{y-(1-\lambda)u}{\lambda}\right)dy$$

Crowder (1987) states that $L(x)$ is a Fredholm integral equation of the second kind, and adds that it can be estimated numerically. The normal probability density function Φ is available in Excel but the Fredholm integral equation is not readily accessible.

Excel can, in fact, handle numerical integration with a Visual Basic for Applications program, and my experience is that Romberg integration (Hornbeck, 1975, 150–154) delivers the best results. This is, however, for single integration only, and solution of the Fredholm integral is apparently far more complicated. Calculation of the ARL for EWMA control charts therefore looks like the only application in this book that is not readily deployable in Excel. StatGraphics can, however, generate an operating characteristic curve and ARL graph for EWMA control charts, and it uses the methods described by Crowder.

Summary

This chapter has shown how to use CUSUM and EWMA as alternatives to traditional SPC for a single quality characteristic. Both methods will detect small discrepancies between the process mean and the nominal more rapidly than traditional SPC, and both are usable for individual measurements or sample averages. The chapter has also provided awareness of Cuscore, an extremely powerful method with which to detect a *specific* kind of process shift such as an increase in tool wear rate from that which is expected.

The next chapter will address applications in which a tool produces parts with different specification limits, but with only one standard deviation. This supports mixed model production, e.g. 5As, 3Bs, and 4Cs to meet just-in-time demand from a downstream process.

Chapter 4

Charts for Multiple Nominals

AIAG (2016, 21–22) cites production of parts with different nominals by the same tool, and adds that DNOM (deviation from nominal) charts for sample averages and ranges work best for constant subgroup sizes. The previous chapter showed, however, that use of standard normal deviates for the sample averages, and quantiles of the chi-square statistic for the sample standard deviations, eliminates the need for constant sample sizes. This is useful if, for example, the downstream process (or kanban container) requires five units of A, three of B, and four of C.

The AIAG reference adds that use of separate control charts for each part is not the best way to assess the status of the process. Another consideration is that it is far less convenient to keep three sets of charts for the sample average and variation than it is to keep one. The approach from the previous chapter is therefore easily adaptable as follows for measurements of n parts of type j. Recall that the chi-square test for sample standard deviation tests the null hypothesis that the sample comes from a population with an assumed standard deviation, $\sigma_{process}$ in this case, versus the alternative hypothesis that it does not.

$$z = \frac{\bar{x} - \text{nominal}_j}{\frac{\sigma_{process}}{\sqrt{n}}}$$

$$\chi^2_{n-1} = \frac{(n-1)s^2}{\sigma^2_{process}}$$

DOI: 10.4324/9781003281061-4

This is simply the procedure that was used for a single part type in the previous chapter. It is, however, mandatory, as pointed out by the AIAG (2016, 23) that the variation for each part type be similar if not identical. If this condition is not met, it is easy enough to modify the formulas shown above to accommodate different variations where σ_j is the process standard deviation for parts or features of type j, and this will be covered later. Equations 4.1 and 4.2 show how to do this.

$$z = \frac{\bar{x} - \textit{nominal}_j}{\dfrac{\sigma_j}{\sqrt{n}}} \tag{4.1}$$

$$\chi^2_{n-1} = \frac{(n-1)s^2}{\sigma_j^2} \tag{4.2}$$

AIAG (2016, 23) describes a test to determine whether we can assume that the part standard deviations are equal. Calculate the ratio, for each part type j,

$\dfrac{\bar{R}_j}{\bar{\bar{R}}}$ where $\bar{R}_j$ = average range for part j, $\bar{\bar{R}}$ = grand average range

If this ratio is less than 1.3, then we can assume that the part variations are in fact the same. Note, however, that if we have enough process history to have average ranges for each part, we may have enough to estimate the process parameters in the manner usual for standard control charts.

Levene's Test and Bartlett's Test also are options. Table 4.1 contains data from four part types with standard deviations of 0.2, 0.3, 0.2, and 0.2, respectively.

StatGraphics obtains the results shown in Figure 4.1 for Levene's Test and Bartlett's Test. In both cases, the P-value is the "reasonable doubt" to the effect that the observed differences are due to random chance rather than a real difference. The traditional cutoff is 0.05 or 5%, i.e. we are 95% sure there is really a difference. In both cases, the initial test statistics with their P-values of 0.0006 and 0.00026, respectively, tell us that the part types do not have the same variation. The comparisons then show clearly that there is no apparent difference between the variations for types A, C, and D, but all of them differ from type B.

StatGraphics also provides a box and whisker plot (Figure 4.2) that shows qualitatively that the characteristic for part type B has more variation than those for A, C, and D. The height of the box reflects the amount of variation.

Table 4.1   Data for Parts with Unequal Variances

A	B	C	D
10.084	10.123	10.151	10.019
9.780	10.549	10.260	9.946
10.137	10.045	9.881	10.148
10.111	9.906	10.129	9.868
9.918	9.763	9.926	10.161
10.382	9.604	9.983	10.026
10.297	10.835	10.221	10.057
9.812	9.334	10.162	9.654
10.055	9.778	9.718	10.054
10.067	10.483	9.844	10.112
10.261	10.174	10.053	10.146
10.271	9.557	10.153	10.051

Variance Check

	Test	P-Value
Levene's	7.06	0.0006

Comparison	Sigma1	Sigma2	F-Ratio	P-Value
A / B	0.18992	0.44603	0.18131	0.0086
A / C	0.18992	0.16864	1.2682	0.7004
A / D	0.18992	0.14355	1.7505	0.3671
B / C	0.44603	0.16864	6.9949	0.0032
B / D	0.44603	0.14355	9.6546	0.0007
C / D	0.16864	0.14355	1.3802	0.6021

Variance Check

	Test	P-Value
Bartlett's	19.074	0.00026396

Comparison	Sigma1	Sigma2	F-Ratio	P-Value
A / B	0.18992	0.44603	0.18131	0.0086
A / C	0.18992	0.16864	1.2682	0.7004
A / D	0.18992	0.14355	1.7505	0.3671
B / C	0.44603	0.16864	6.9949	0.0032
B / D	0.44603	0.14355	9.6546	0.0007
C / D	0.16864	0.14355	1.3802	0.6021

Figure 4.1   Levene's Test and Bartlett's Test in StatGraphics

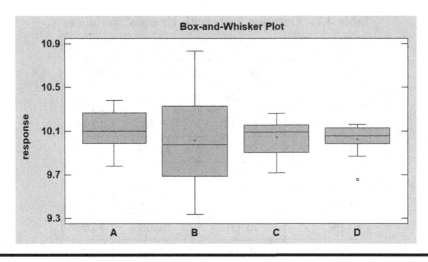

Figure 4.2 Box and whisker plot

In addition, if we have only two part types, we can also use the F test for equality of variances. The takeaway so far is that (1) this form of short-run SPC for parts with different nominals and equal variances requires assessment of the assumption that the variances are equal and (2) there are numerous quantitative and qualitative (graphical) ways to do this. *All these tests assume, however, that we have historical data from which we can calculate the average range for each part and the grand average range.* If there are no historical data or surrogate process, then we must infer the standard deviations from (1) the specification width and (2) the required process performance index as before. If the specification widths are not identical, then we should probably assume that the part standard deviations are not identical either.

Parts with Different Nominals

Given kanban containers that hold 2 A, 4 B, and 3 C with nominals of 4.00, 4.50, and 5.00 mm, respectively. The specification width in each case is 0.1 mm and the required process performance index is 5/3. Recall from Equation 2.1,

$$\sigma_{\text{process}} = \frac{\text{USL} - \text{LSL}}{6P_{\text{p}}} = \frac{0.1\,\text{mm}}{6\dfrac{5}{3}} = 0.01\,\text{mm}$$

Equation 4.3 delivers the standard normal deviate for each part, noting that A comes in groups of two, B in groups of four, and C in groups of three:

$$z = \frac{dnom}{\dfrac{\sigma_{\text{process}}}{\sqrt{n}}} = \frac{\bar{x} - nominal}{\dfrac{\sigma_{\text{process}}}{\sqrt{n}}} \tag{4.3}$$

$$z_A = \frac{\left(\bar{x} - 4.0\right)\text{mm}}{\dfrac{0.01\,\text{mm}}{\sqrt{2}}} \qquad z_B = \frac{\left(\bar{x} - 4.5\right)\text{mm}}{\dfrac{0.01\,\text{mm}}{\sqrt{4}}} \qquad z_C = \frac{\left(\bar{x} - 5.0\right)\text{mm}}{\dfrac{0.01\,\text{mm}}{\sqrt{3}}}$$

Equation 4.4 delivers the chi-square test statistic for each sample, where s is the sample standard deviation. The latter is in millimeters, so the variance is in square millimeters. The chi-square statistics for A, B, and C have 1, 3, and 2 degrees of freedom, respectively, i.e. one less than the sample size in each case.

$$\chi^2_{n-1} = \frac{(n-1)s^2}{\sigma_0^2} = \frac{(n-1)s^2}{\sigma_{\text{process}}^2} \tag{4.4}$$

$$A: \chi_1^2 = \frac{(2-1)s^2\,\text{mm}^2}{0.01^2\,\text{mm}^2} \qquad B: \chi_3^2 = \frac{(4-1)s^2\,\text{mm}^2}{0.01^2\,\text{mm}^2} \qquad C: \chi_2^2 = \frac{(3-1)s^2\,\text{mm}^2}{0.01^2\,\text{mm}^2}$$

Assume that the process standard deviation does meet the 0.01 mm requirement, but C is running below nominal at 4.98 mm. Five lots were simulated with the results shown in Figure 4.3.

Consider, for example, Lot 1 and Part B, for which there are four measurements. The standard normal deviate is, where VLOOKUP looks up the correct nominal from cell B13 (which contains B as the part) from the table in cells A7 through B9. The 2 as the last argument says to extract the value from the second column, which contains the nominals. The ISBLANK test prevents the formula from acting on a row in which there are no measurements.

=IF(ISBLANK(C13),NA(),(AVERAGE(C13:F13)-VLOOKUP(B13,A$7:B$9,2))/(B$5/SQRT(COUNT(C13:F13))))

$$z_B = \frac{(4.495 - 4.5)\,\text{mm}}{\dfrac{0.01\,\text{mm}}{\sqrt{4}}} = -1.00$$

	A	B	C	D	E	F	G	H	I
1	Charts for Parts with Different Nominals, Same Variation					2-tailed false alarm risk for chi square			
2						alpha	0.0027	User entry	
3	Specification Width		0.1	user entry		LCL	0.00135	Calculated	
4	Required Pp	1.666667		user entry		UCL	0.99865	Calculated	
5	sigma_process		0.01	calculated					
6	Part nominals	nominal							
7	A		4.0	user entry					
8	B		4.5	user entry					
9	C		5.0	user entry					
10									
11	Description	Part	X1	X2	X3	X4	z	Chi square	quantile
12	Lot 1	A	4.011	4.000			0.77	0.588	0.557
13	Lot 1	B	4.495	4.489	4.482	4.515	-1.00	6.101	0.893
14	Lot 1	C	4.990	4.987	4.990		-1.89	0.051	0.025
15	Lot 2	A	3.991	3.995			-0.95	0.092	0.238
16	Lot 2	B	4.500	4.490	4.484	4.501	-1.27	1.975	0.422
17	Lot 2	C	4.973	4.973	4.981		-4.18	0.359	0.164
18	Lot 3	A	4.009	4.008			1.14	0.005	0.057
19	Lot 3	B	4.484	4.504	4.508	4.500	-0.20	3.303	0.653
20	Lot 3	C	4.973	5.000	4.975		-2.98	4.721	0.906
21	Lot 4	A	4.011	3.997			0.52	1.028	0.689
22	Lot 4	B	4.510	4.503	4.510	4.481	0.21	5.513	0.862
23	Lot 4	C	4.985	4.986	4.988		-2.38	0.050	0.025
24	Lot 5	A	4.009	3.994			0.26	1.162	0.719
25	Lot 5	B	4.506	4.501	4.506	4.512	1.25	0.579	0.099
26	Lot 5	C	4.982	4.975	4.979		-3.74	0.238	0.112

Figure 4.3   Control spreadsheet for parts with different nominals, same variation

The cell for the standard normal deviate will turn red if the value is less than −3 or greater than 3. The chi-square statistic is:

=IF(COUNT(C13:F13)<2,NA(),((COUNT(C13:F13)-1)*STDEV(C13:F13)^2/B$5^2))

$$\chi^2 = \frac{(4-1)0.01426^2 \text{ mm}^2}{0.01^2 \text{ mm}^2} = 6.10$$

The cell for its quantile will turn green if the quantile is less than the specified false alarm risk, 0.027 in this case divided by 2. It will turn red if the quantile is greater than 1 minus the specified false alarm risk divided by 2. Figure 4.4 shows how the control charts might be presented in Excel. The chart for the standard normal deviate reinforces the conclusion that part C is running well below its specified nominal.

Remember again that the chart for the quantiles of the chi-square statistic is not a classical Shewhart-type chart in which we expect most of the points to concentrate within 1 standard deviation of the center line, as the quantiles

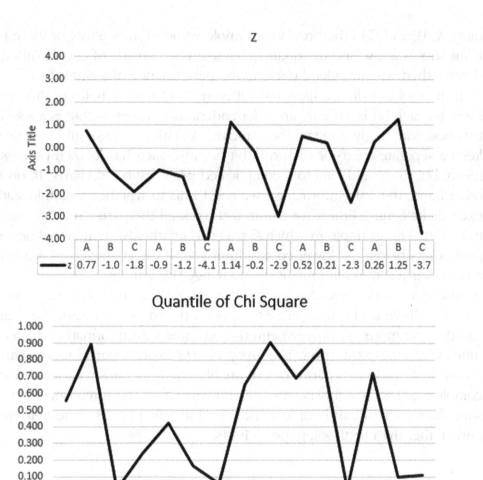

	A	B	C	A	B	C	A	B	C	A	B	C	A	B	C
z	0.77	-1.0	-1.8	-0.9	-1.2	-4.1	1.14	-0.2	-2.9	0.52	0.21	-2.3	0.26	1.25	-3.7

Figure 4.4 Control charts for different nominals, same variance

follow the uniform distribution. The standardized range chart or standard deviation chart can be used instead of the quantile of the chi-square statistic if this is preferred, although the control limits will then vary according to the sample size.

Practical Consideration: Number of "Tools"

It is rarely, if ever, a good idea to track the performance of more than one tool on any given control chart, especially if the chart uses the Western Electric zone tests and/or the user looks for trends and patterns as well as out of control signals. This means we must address the question as to whether parts A, B, and C in the previous example come from one tool or three tools. If the tool is a saw and the quality characteristic consists of cuts of different depths, then we are indeed tracking the performance of a single tool.

If the tool is a drill, it must use different bits to make holes of different sizes, so each bit is actually an independent tool whose output is a size distribution with its own mean and standard deviation.* We should technically keep a separate chart for each drill bit because each has its own process mean, but this could lead to a complicated proliferation of charts. If, on the other hand, the only information we need is as to whether a sample statistic exceeds its control limits, we can in fact track all the parts on the same control chart. The example in which C was running below its nominal generated points below the lower control limit and, if the characteristic was in fact a hole, our reaction plan might be to change the drill bit.

The same issue arises for parts with multiple characteristics. Each bit of a multi-spindle drill is a separate tool, but a dozen or more separate charts can do little more than overwhelm the operator with information that takes time away from production. The *group chart's* express purpose is to handle a very wide variety of part types or quality characteristics without excessive complexity. The key consideration is, however, that the spreadsheet is the shop floor's servant and not its master, and its sole purpose is to help make parts rather than textbook-perfect charts.

Parts with Multiple Characteristics, Same Variation

AIAG (2016, 25) provides an example of a valve body in which a CNC (computer numerical control) drills 12 holes: four pressure holes whose nominal is 4.5 mm, four CW (clockwise turn from context) holes whose nominal is 2.75 mm, and four CCW (counterclockwise turn) holes whose nominal is 2.65 mm. The specification width for each feature is 0.05 mm, which makes it reasonable to believe that the variation is the same for each hole, noting especially that the same machine and spindle perform the job. This can be handled almost identically to the situation for different parts with different nominals and the same variation. There is also no requirement that the number of holes in each category be equal.

---

* The question as to whether the hole sizes follow a normal distribution is another question entirely, as the hole diameter cannot be less than that of the drill bit. Not much information is apparently available on the Internet on the probability density functions of typical machining operations, although the normal distribution seems to be "good enough" in most situations.

Suppose instead that a part has three characteristics of type A whose nominal is 5.000 mm, five of B whose nominal is 5.500 mm, and four of C whose nominal is 6.000 mm. The specification width is 0.060 mm in all cases and the required process performance index is 4/3, i.e. a 4-sigma process.

$$\sigma_{process} = \frac{0.060\,mm}{6\frac{4}{3}} = 0.0075\,mm$$

Figure 4.5 shows how to deploy this in a spreadsheet. The out of control signal for Part 5, characteristic C, is in fact due to random variation as no assignable cause was simulated.

Cell H12 is =IF(ISBLANK(C12),NA(),(AVERAGE(C12:G12)-VLOOKUP(B12, A\$7:B\$9,2))/(B\$5/SQRT(COUNT(C12:G12)))) which divides the average of the measurements for X1 through X5 minus the nominal, while ignoring blanks, by the specified process standard deviation after division of the latter by the

	A	B	C	D	E	F	G	H	I	J
1	Charts for Parts with Multiple Characteristics, Same Variation					2-tailed false alarm risk for chi square				
2						alpha	0.0027	User entry		
3	Specification Width	0.06	user entry			LCL	0.00135	Calculated		
4	Required Pp	1.333333	user entry			UCL	0.99865	Calculated		
5	sigma_process	0.0075	calculated							
6	Feature	nominal								
7	A		5.0	user entry						
8	B		5.5	user entry						
9	C		6.0	user entry						
10										
11	Part		Feature	X1	X2	X3	X4	X5	z	Chi square	quantile
12	1		A	5.000	5.010	4.991			0.10	3.001	0.777
13	1		B	5.510	5.500	5.499	5.499	5.503	0.63	1.523	0.177
14	1		C	5.997	6.005	6.015	5.992		0.55	5.416	0.856
15	2		A	5.001	5.000	5.008			0.71	0.738	0.309
16	2		B	5.504	5.502	5.497	5.504	5.497	0.26	0.822	0.065
17	2		C	6.004	6.000	5.997	5.990		-0.58	1.745	0.373
18	3		A	4.994	5.001	5.003			-0.19	0.873	0.354
19	3		B	5.501	5.499	5.506	5.505	5.504	0.88	0.580	0.035
20	3		C	5.987	6.001	6.004	6.006		-0.16	3.952	0.733
21	4		A	5.001	5.003	5.000			0.32	0.049	0.024
22	4		B	5.508	5.504	5.491	5.507	5.494	0.26	4.283	0.631
23	4		C	6.006	6.000	6.001	5.994		0.08	1.430	0.302
24	5		A	4.997	5.000	5.006			0.19	0.722	0.303
25	5		B	5.496	5.485	5.510	5.506	5.491	-0.62	7.664	0.895
26	5		C	6.023	6.020	6.011	6.003		3.72	4.421	0.781
27										

Figure 4.5   Spreadsheet for parts with multiple characteristics

square root of the number of measurements. As an example, for part 5 and characteristic C; the result differs slightly because the spreadsheet carries all the significant figures.

$$z = \frac{6.014 - 6.000\,\text{mm}}{\dfrac{0.075\,\text{mm}}{\sqrt{4}}} = 3.73$$

The resulting charts can be displayed as shown in Figure 4.6. It is necessary to use the Quick Layout feature in Excel to get the axis labels on the bottom of the chart rather than in the middle.

Note again that the production worker needs to do no more than enter the part number, characteristic, and measurements to get actionable information.

Parts with Multiple Characteristics and Different Variations

Suppose we add the complication that the characteristics have not only different nominals but also different amounts of variation even though the same tool produces them. Equations 4.3 and 4.4 (2.2 and 2.3 from Chapter 2) already show how to address this situation, in this case for characteristic j when the sample size is n. While we can use a standardized sample range or sample standard deviation chart for the variation instead, use of the chi-square statistic eliminates the need for the spreadsheet to look up the control chart factor.

Group Charts

The previous chapter addressed SPC for parts or quality characteristics with different specifications, and even different amounts of variation. Different specifications are addressed easily through conversion of sample averages into standard normal deviates, while sample standard deviations can be converted similarly into chi-square test statistics with known quantiles. Standardized range or standard deviation charts are alternatively feasible for users who wish to stay with traditional methods, although it is then necessary to include in the spreadsheet a table of control chart factors.

A practical problem arises, however, when it is necessary to track a large number of part types and/or quality characteristics. The example with three different parts required three axis points per lot, and the one with three different characteristics required three axis points per part. It is easy to

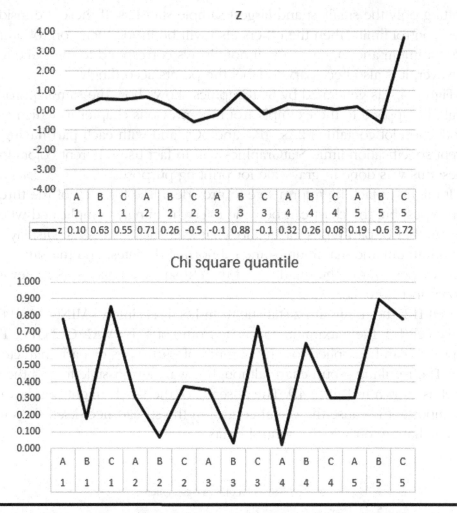

Figure 4.6   z Chart and quantiles of chi-square statistic, multiple product features

envision the charts becoming unwieldy as the number of specifications and/ or characteristics increases, although the complexity of the visual controls built into the spreadsheet cells will not change. The *group chart* displays only one set of points per production lot of different parts, or per part with multiple characteristics.

The Group Chart

Wise and Fair (1998, 177) offer an example in which a machined shaft is measured at its ends and also in its middle. A sample size of four is used in which four shafts are measured at the indicated positions, and sample averages and ranges are then plotted for the three positions. There is only one axis point for each sample of four and further simplification is achieved by plotting only the smallest and largest sample statistics. If these are inside their control limits, then the others also will be inside their control limits. As the minimum and maximum will not always correspond to the same feature, however, it is also necessary to label the points accordingly.

Figure 4.7 as generated by StatGraphics shows how the same approach might be applied to the example from the previous chapter in which each production lot contained 2As, 4Bs, and 3Cs, and with each part having a different specification limit. StatGraphics will in fact use different colors for the lines; this was done in grayscale for printing purposes.

It might, in fact, be informative to provide a line for each of the three part types, but the chart can become overwhelmingly complicated when the number of part types (or quality characteristics) increases. Display of the minimum and maximum standard normal deviates, and the same principle carries over to the quantiles of the chi-square statistics, is achievable as shown in Figure 4.8.

Cell I12, the minimum z statistic from Lot 1, is simply=MIN(G12:G14) while cell J13, the maximum z statistic from Lot 1, is=MAX(G12:G14). This approach can be applied for any number of part types or part characteristics. The resulting group chart, although it was not possible to add the part label, is shown in Figure 4.9. The absence of the labels might not, however, be important because the visual control in the spreadsheet itself flags the points that are outside the control limits.

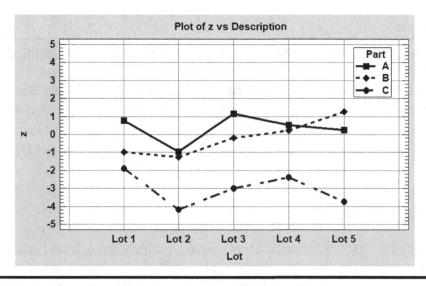

Figure 4.7   Group z chart for lots with parts of different nominals

▲	A	B	C	D	E	F	G	H	I	J
1	Charts for Parts with Different Nominals, Same Variation					2-tailed false alarm risk for chi square				
2						alpha	0.0027	User entry		
3	Specification Width		0.1	user entry		LCL	0.00135	Calculated		
4	Required Pp	1.666667		user entry		UCL	0.99865	Calculated		
5	sigma_process		0.01	calculated						
6	Part nominals	nominal								
7	A		4.0	user entry						
8	B		4.5	user entry						
9	C		5.0	user entry						
10										
11	Description	Part	X1	X2	X3	X4	z	Lot	Min_z	Max_z
12	Lot 1	A	4.011	4.000			0.77	Lot 1	-1.892	0.77
13	Lot 1	B	4.495	4.489	4.482	4.515	-1.00			
14	Lot 1	C	4.990	4.987	4.990		-1.89			
15	Lot 2	A	3.991	3.995			-0.95	Lot 2	-4.179	-0.95
16	Lot 2	B	4.500	4.490	4.484	4.501	-1.27			
17	Lot 2	C	4.973	4.973	4.981		-4.18			
18	Lot 3	A	4.009	4.008			1.14	Lot 3	-2.980	1.14
19	Lot 3	B	4.484	4.504	4.508	4.500	-0.20			
20	Lot 3	C	4.973	5.000	4.975		-2.98			
21	Lot 4	A	4.011	3.997			0.52	Lot 4	-2.376	0.52
22	Lot 4	B	4.510	4.503	4.510	4.481	0.21			
23	Lot 4	C	4.985	4.986	4.988		-2.38			
24	Lot 5	A	4.009	3.994			0.26	Lot 5	-3.738	1.25
25	Lot 5	B	4.506	4.501	4.506	4.512	1.25			
26	Lot 5	C	4.982	4.975	4.979		-3.74			

Figure 4.8   Smallest and largest z statistics from three or more part types

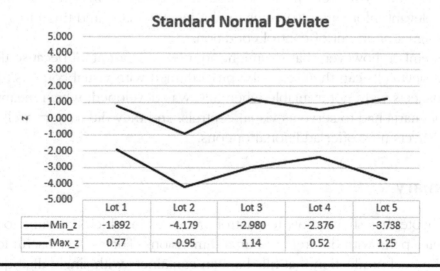

Figure 4.9   Group chart for smallest and largest z statistics

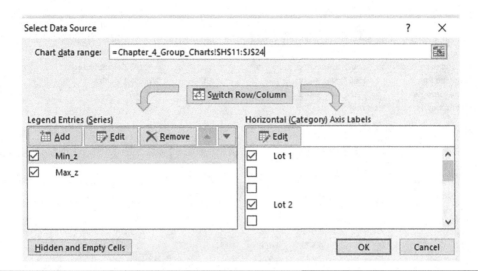

Figure 4.10 Use of select data source to eliminate empty rows from the chart

Note also that the user will have to use the Select Data feature (Figure 4.10) to remove the rows that contain no data to avoid having them show up in the chart. The user must uncheck the boxes for the rows that contain no data. The chart format that places all the labels at the bottom can be obtained from the Quick Layout panel.

The setup of the group chart is somewhat more work for the process owner, who must add the additional columns for the information to be charted, and also set up the proper layout of the chart itself, but the production worker still needs to do no more than enter the measurements and other relevant information (such as the lot identity, date, and time) to get a visual signal of an out of control condition.

Remember, however, that we might not need a chart at all because the spreadsheet cells can themselves be programmed with visual controls. Spreadsheets were not available when SPC was developed, which meant control charts had to serve as visual controls, and they did so very well, but spreadsheets now offer additional options.

Summary

This chapter has shown how to control processes in which the same tool generates parts with different nominal dimensions. This is very useful for mixed-model production that fills kanban containers with single-digit quantities of different parts. It has also shown how to control processes in which a tool makes different product features with different nominal sizes and even different amounts of process variation.

Chapter 5

Acceptance Control Charts

Acceptance *control charts* relate to the process rather than products, which is the context of *acceptance sampling*. Suppose we have a process in which it is not practical or not desirable to attempt to control it to the nominal. Harriott (1964, 6–7) gives the example of an on–off temperature controller for a chemical process kettle in which there is a differential around the set point to avoid wear on the controller. That is, were the controller to act on any deviation from the set point, it would cycle on and off continuously. The result is, depending on the speed at which the system reacts to the control action, a triangular wave or a sine wave. This exemplifies a system in which the output cannot be held to the target, set point, or nominal but can be held between quantifiable boundaries. Box and Luceno (1997, 147–149) define a *dead band* as a region around the target (or nominal) in which no control action is desirable. One benefit of a dead band is a reduction in adjustment actions. This can be helpful if there is a time or money cost related to adjustments, especially if they are manual.

Other applications exist in which it is not possible and/or desirable to hold a process to a set point or nominal, but we want to hold the mean to some level inside a set of boundaries. These are known as the lower acceptable process level (LAPL) and upper acceptable process level (UAPL). Feigenbaum (1991, 426–428) discusses acceptance control charts in further detail.

The process owner can set LAPL and UAPL as simply the boundaries of the region in which it is practical and desirable to hold the process mean. Feigenbaum (1991, 427) defines an alternative method in which the process owner sets these to reflect the acceptable nonconforming fraction. If this fraction is p1, then Equation 5.1 shows how to calculate these, where LSL and USL are the lower and upper specification (or tolerance) limits. If we have no prior

DOI: 10.4324/9781003281061-5

information about the process, then sigma is defined as it was previously, e.g. based on the required process performance index. The region between LAPL and UAPL is known as the *indifference zone*, and may be called alternatively a center band (as opposed to a center line) or a dead band.

$$LAPL = LSL + z_{p1}\sigma \quad UAPL = USL - z_{p1}\sigma \tag{5.1}$$

The familiar "3.4 million defects per million opportunities" relates to a Six Sigma process with a 1.5-sigma process shift. If we have a Six Sigma process and are willing to accept 3.4 DPMO, we would then use a standard normal deviate of 1.5. The width of the center band or indifference zone would then be 3 standard deviations.

Calculation of the control limits (or acceptance control limits) is then very straightforward, as shown in Equation (5.2), where alpha is the desired false alarm risk of going over the control limit if the process mean is at the acceptable process level. The risk for the traditional Shewhart chart is 0.00135 for which the standard normal deviate z is 3.

$$LCL = LAPL - \frac{z_\alpha \sigma}{\sqrt{n}} \quad UCL = UAPL + \frac{z_\alpha \sigma}{\sqrt{n}} \tag{5.2}$$

The easiest way to deploy this on a spreadsheet is to set the limits as shown in Equation 5.2 but it might be simpler to use wider control limits. Suppose, for example, we have a 5-sigma process and are willing to tolerate a 1-sigma process shift. The control limits can then be set at the nominal $\pm$4-sigma rather than $\pm$3-sigma.

Use of the Western Electric zone tests, e.g. the Zone C test that calls eight consecutive points above or below the center line evidence of a process shift, would, however, add a complication. An acceptance sampling chart that seeks to use this test would use instead eight consecutive points above the UAPL or below the LAPL, with those in between not counting.

Equation 5.3 modifies Equation 5.2 to plot sample standard normal deviates to allow different sample sizes, and also uses the LAPL and UAPL as opposed to merely wider control limits.

$$z_{\bar{x}} = \frac{\bar{x} - \text{nominal}}{\dfrac{\sigma}{\sqrt{n}}} \tag{5.3a}$$

$$LCL = \frac{LAPL - \text{nominal}}{\sigma} - z_\alpha \quad UCL = \frac{UAPL - \text{nominal}}{\sigma} + z_\alpha \tag{5.3b}$$

Twenty-four samples of different sizes (2 and 3) were simulated from a Six Sigma process with specifications [99,101] and a mean of 100.1. The process standard deviation is therefore 1/6 but the process is off center by 0.1. We are willing to accept a process shift of 1-sigma, i.e. any process mean between 99.83 and 100.17. The lower control limit is therefore simply −4 and the upper control limit is 4, as shown in Figure 5.1, as generated by StatGraphics. (The values are from the Excel spreadsheet that accompanies this book, but StatGraphics creates a better plot.)

If we were using instead a traditional SPC chart with a center line of 0 for the sample average's standard normal deviate, and control limits of ±3, the chart looks like it would fail the Zone A test for two out of three consecutive points more than two standard deviations above the center line. As we are willing to accept a 1-sigma process shift, however, the points in question are in Zone B (four out of five consecutive points more than 1 standard deviation above or below the center line, and there does not appear to be any failure of this test) rather than Zone A.

There is, however, no difference in the way the charts for process variation work. We can, for example, use the chi-square statistic (or its quantile) to identify out of control situations for samples of different sizes; these are functions of the sample standard deviation. We can alternatively use traditional control limits for the sample range or standard deviation that depend on the sample size.

In summary, then, the acceptance control chart is a good practical alternative when the process owner does not find it desirable and/or practical to hold the process to the nominal.

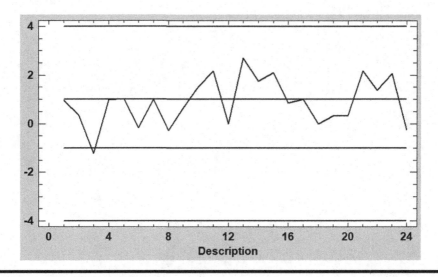

Figure 5.1   Acceptance z chart

Bibliography

ASTM. 1990. *Manual on Presentation of Data and Control Chart Analysis.* 6th edition. Philadelphia, PA: American Society for Testing and Materials.

AT&T. 1985. *Statistical Quality Control Handbook.* Indianapolis, IN: AT&T Technologies.

Automotive Industry Action Group. (AIAG). 2016. *SPC Short Run Supplement.* Southfield, MI: AIAG.

Automotive Industry Action Group. (AIAG). 2005. *Statistical Process Control.* 2nd edition. Southfield, MI: AIAG.

Box, George, and Luceño, Alberto. 1997. *Statistical Control by Monitoring and Adjustment Feedback.* New York: John Wiley & Sons.

Box, George, and Ramirez, Jose. 1992. "Cumulative Score Charts." *Quality and Reliability Engineering International,* 8, 17–27.

Crowder, Stephen. 1987. "A Simple Method for Studying Run – Length Distributions of Exponentially Weighted Moving Average Charts." *Technometrics,* 29:4, 401–407.

Crowder, Stephen. 1989. "Design of Exponentially Weighted Moving Average Schemes." *Journal of Quality Technology,* 21:3, 155–162.

Feigenbaum, Armand V. 1991. *Total Quality Control.* 3rd edition. New York: McGraw-Hill.

Harriott, Peter. 1964. *Process Control.* New York: McGraw-Hill.

Heinlein, Robert A. 1959. *Starship Troopers.* New York: G. P. Putnam's Sons.

Hornbeck, Robert W. 1975. *Numerical Methods.* New York: Quantum Publishers.

Juran, Joseph, and Gryna, Frank. 1988. *Juran's Quality Control Handbook.* New York: McGraw-Hill.

Maghsoodloo, S. 2013. "Cumulative Sum (CUSUM) Control Charts." https://www.eng.auburn.edu/~maghssa/INSY7330/Cusum-Control-Charts-M2013-Maghsoodloo.pdf.

Montgomery, Douglas. 1991. *Introduction to Statistical Quality Control.* 2nd edition. New York: John Wiley & Sons.

Smith, Jim. 2009. "PRE-Control May Be the Solution." *Quality Magazine.* https://www.qualitymag.com/articles/86794-pre-control-may-be-the-solution.

Wise, Stephen, and Fair, Douglas. 1998. *Innovative Control Charting.* Milwaukee: ASQ Quality Press.

Index